# LEVEL UP

## Middle Grades Math Edition

Edited by Matthew Deevers, PhD
Summit Education Initiative Senior Research Associate



Inspirations and Sponsorship Provided by

GOODYEAR

The Goodyear Tire & Rubber Company

**Level UP**
Middle Grades Math Edition

FIRST EDITION

ISBN: 978-0-9978068-1-6

Library of Congress Control Number: 2018900442



Edited by Matthew Deevers, PhD

Published by Summit Education Initiative
120 E. Mill Street, Suite 330
Akron, Ohio 44308
www.seisummit.org

Summit Education Initiative

# Acknowledgements

Summit Education Initiative would like to thank The Goodyear Tire & Rubber Company® for its financial and operational support for Level UP: Middle Grades Math Edition. In particular, Maureen Thune and Alison Lauge offered leadership, guidance and support throughout the process. Many Goodyear associates graciously shared personal experiences with persisting through challenges, fitting in and finding their way in life. These stories became the inspirations used throughout the book; each contributing associate is credited for her or his story when it appears.

Amy Kamlowsky provided math problems for the book, along with hints and solutions to support students. Michelle Burrier offered significant support in the organization and formatting of this book.

- Summit Education Initiative, 2018

This page intentionally left blank

# Table of Contents

# Introduction

## *Hey. We're talking to you. Yeah, you!*

This is a math book for *you*.

Did you think it was for someone else? Nope. It's for you.

Get over it.

No matter who you are, there is something for you in this book.

Maybe one of the stories in this book will inspire you. Maybe you will learn something about yourself in a "reflection zone." Maybe you'll complete a set of challenge problems and be more confident. Or maybe you'll read one of the fun facts in the book and share it with a friend.

We promise you will find at least one part of this book that you like.

And you know what else? We promise you will be a stronger math student at the end of this book. That's because we're going to give you lots of practice.

Everyone can be better at math.

Math isn't something you're born with. Your brain gets stronger every time to work through a new challenge.

Math is something that takes attention, practice and a whole lot of mistakes. If math was supposed to be easy, why did they put erasers on pencils?!?

So do yourself a favor: use this book. Read the stories. Write down your reflections. Do the practice challenges. Set goals. Your brain will thank you.

# How to use this book

This book is written for all kinds of middle grade math students. There are 4 levels of math problems you can work through. Level 1 is the easiest level; level 4 is the hardest level. Your teacher might assign a level to you; if not, you can look at the problems and decide which is the best challenge for you.

There are two practice tests in this book. We recommend that you take Practice Test One before you do anything else. After you take Practice Test One, work through the book from the beginning.

There are 8 chapters in the book. Each chapter has a few personal stories from adults who work at The Goodyear Tire & Rubber Company's international headquarters in Akron, Ohio. After each story, there is a spot for you to write down a reflection. *Reflection is an important part of this book.*

After you read the stories and write the reflections in a chapter, you will see the challenge problems. Take your time and work the challenge problems for your level. There are hints at the back of the book if you need a little help. Work through the self-assessment zone after the challenge problems. This will help you know how you're doing, and what you need to work on. Then set a goal for yourself in the next chapter and move on.

Take Practice Test Two after you finish Chapter 8. That will help you see how much you have improved!

Think about this book like exercise for your brain. The more you use it, the stronger your brain will get. When you do all the exercises, your brain will definitely **Level UP!**

# More stuff about this book

We already said this, but there are 8 chapters in this book. You should work through one chapter each week. We also broke up the practice problems into different kinds of math. You will practice one kind of math each week, and you will see each kind of math *twice*. This is how it works:

| Chapter | Challenge Problems |
|---|---|
| One | Algebra |
| Two | Geometry |
| Three | Number Sense |
| Four | Data & Statistics |
| Five | Algebra |
| Six | Geometry |
| Seven | Number Sense |
| Eight | Data & Statistics |

## Use the self-assessment zones!

You will learn more and get stronger if you use the self-assessments after your challenge problems. Take time to do this, and *be honest.* If you got the right answer, but you had to guess, mark that down. You will get to know yourself better, and you will know how to improve.

## Set goals!

The most successful people in the world write down their goals, and then work to achieve them. Take some time to set a goal for yourself after each chapter. Think about what you need to do in order to reach that goal. Every time you set a goal and work toward it, you get closer to success.

It is time to

# LEVEL UP

Your Middle Grades Math Skills

# LET'S DO THIS!

# Chapter 1

## No such thing as The Easy Road

You may have heard that *the road to success is paved with hard work*. Did you also know that road can have big bumps, accidents and detours?

It's true.

The authors of these stories all learned they needed to work hard to get where they wanted in life. When you read these stories, think about how you have worked to reach your goals.

Also, think about how hard you are willing to work to reach your *next* goal.

**Robert Outzen**
**Goodyear Employee Starting in 1985**

## From Mississippi to Manager

*Read Robert's story to learn how he worked to reach his career goal, even after multiple failures*

I was raised in Mississippi. My best friend was murdered just before my 18th birthday. My father passed away four months later.

I completed high school and attempted to go to college, but I failed miserably. I dropped out and began working on the Mississippi River as a deckhand. A few months into the job, I was almost blown up with the ship. That ended my career on the water!

In the meantime, my mother had moved the family to Florida and encouraged me to move, too; which I did. Upon arriving in Florida, I began working full-time and attending college, but shortly afterwards I dropped out yet again.

> *"Disappointment became determination, and I obtained my associate degree."*

Fast forward to the 1980s. I started with Goodyear Retail as a salesman. During my first year, the store was audited. After talking with the auditor, I decided that I wanted his job, but I was informed the job required a college degree. After failing twice at college, I was dismayed, but I knew what I had to do to be considered for the auditor position.

Disappointment became determination and I obtained my associate degree while attending night school. Later I learned the position required a bachelor's degree. So even though I succeeded, it still wasn't enough!

In 2001, with the encouragement and support of my wife, I graduated college with a 4.0 GPA while working for Goodyear's commercial division. In 2006, I attained my initial goal when I applied and was accepted into

Goodyear's Field Audit department. In my tenure as auditor, I have continued to learn and accept challenges and opportunities, which has lead me to become manager of the Field Audit department.

Failure and hardship are just part of life. The key is to realize that if you don't succeed at first, keep trying, because you can accomplish anything.

**FUN FACT:** Robert wasn't the only person who started out working on a river boat. Mark Twain, one of America's most important authors, also worked on the Mississippi River before he started writing.

# REFLECTION ZONE

*Write down one goal you have, something you want to accomplish that seems difficult right now.*

**Ricky Brown, Jr.**
**Goodyear Employee Starting in 2003**

# Climbing Through Bad Luck – and Lots of School

*Ricky Brown, Jr. got "lucky" after a really bad accident. Read how Ricky kept his focus through ups and downs*

Not all children are raised by their mother and father. Some are raised in non-traditional settings, by foster parents or even by other family members.

I was raised by my grandparents. This was not the "ideal" childhood, but it was the best option available. I struggled a lot in grade school and did not perform well. I had many favorite subjects, but the "schooling" life never came naturally to me.

I did not go straight into college after high school. Instead, I started working full-time for Goodyear. I spent 10 years climbing the ladder of success. Despite my accomplishments, I knew that I wanted to grow and better myself for my family and Goodyear. I knew it was time to earn my college degree.

Before I was able to go on to college, I was involved in a serious accident. While I was undergoing treatment, my doctors discovered I had a congenital heart disorder that required immediate surgery or else I would not survive. It is hard to believe, but that accident *saved my life*!

Over the next four years, I endured a number of surgeries, countless therapy sessions and doctors' visits – all while working full-time and going to college. I ultimately completed my undergraduate education and graduated with honors.

I am still employed with Goodyear, growing professionally and personally. After all this I now find myself back in school as a graduate student!

*"In my time I have learned that success never comes easy."*

In my time I have learned that success never comes easy. I was not dealt an ideal hand. Life is hard work, but there is much beauty to it. Much of my success comes from the support of my family, loved ones and my self-determination to succeed.

No matter your past or your present state, you can achieve greatness; every accomplishment starts with the decision to begin.

**FUN FACT:** Ricky was raised by his grandparents. So was President Barack Obama, who moved in with his grandparents at the age of 10. Where you come from isn't as important as where you are willing to go.

# REFLECTION ZONE

*Think about your "Decision to begin." Write down one way you can stay focused on your goals.*

**Eric Reckenbeil**
**Goodyear Employee Starting in 2006**

## Learn from Your Successes, Too

*Eric learned to push himself if we wanted to improve. That lesson from sports has helped in life.*

Never give up.

I loved to play basketball for fun as a kid. As I grew up, I was a huge New York Knicks fan and was always impressed by professional players' skills and body control.

I started playing on a basketball team in fifth grade, and I wasn't very good. I never started and was always the last person to be put in the game. But because I liked the game, I continued to practice and began to get better. I would play as much as I could in summer leagues and different pick-up games in the neighborhood. I continued to grow and get better. I would always try to play the older kids because they were much better than me. They challenged my abilities.

In eighth grade, I played in two leagues and I was making more progress on skills. When I got to high school, I tried out and made the freshmen basketball team. I had a great coach who helped me improve even more.

By the time sophomore year rolled around I was starting on the junior varsity team and playing some varsity games. If I had given up when success did not come early, I would have never had the opportunity to play varsity basketball with my high school. By the time I was a senior, I had made "all-county."

> *I would always try to play the older kids because they were much better than me. They challenged my abilities.*

Even though this success was in sports, it gave me confidence to pursue other difficult challenges. I know that if I want something, I must focus and practice long enough to achieve it. This has carried over to my personal and professional lives, where I continue to develop new skills and abilities to become a better person.

**FUN FACT:** Eric isn't the only person who learned a lot from playing sports in high school. Students who play sports or get involved in other school activities in high school, are more successful in life.

# REFLECTION ZONE

*Everyone is good at something. Write down something you already do well. How can you use that skill to motivate yourself in other parts of life?*

# Feeling inspired?

# LEVEL UP

**TRY THESE PRACTICE PROBLEMS**

Each level is a little more challenging

Check your work at the end

Look for hints if you need them

## CHALLENGE SET 1: ALGEBRA

### Level 1 Algebra

1. Evaluate the expression $(9 - 6) \div 3$

2. What is the numerical expression for *Three times the sum of five and six*

3. Fill the table for the value of the function $\boldsymbol{a = b + 8}$

| A | B |
|---|---|
| 8 | 0 |
|  | 1 |
| 10 | 2 |
|  | 5 |

4. Find the missing numbers and determine the rule

| IN | OUT |
|---|---|
| 2 | 4 |
| | 8 |
| 8 | 16 |
| 10 | |

The rule is ______________________________

| IN | OUT |
|---|---|
| 12 | 19 |
| 23 | |
| | 37 |
| 41 | 48 |

The rule is ______________________________

5. Kathy can buy a spool with 10 yards of ribbon at the craft store for $8. If the cost is the same to buy the ribbon per yard, how much will she pay for 5 yards of ribbon?

Need some hints? Look on page 176 for help

## Level 2 Algebra

1. Match the exponent to the equivalent expression

| | |
|---|---|
| A. $2^6$ | 1. 2 x 2 x 2 x 2 x 2 |
| B. $5^3$ | 2. 7 x 7 x 7 x 7 x 7 x 7 x 7 |
| C. $2^5$ | 3. 2 x 2 x 2 x 2 x 2 x 2 |
| D. $5^2$ | 4. 5 x 5 |
| E. $7^7$ | 5. 5 x 5 x 5 |

2. Alex has 12 free downloads for music. His uncle bought him *s* more songs. Write an expression to show how many downloads Alex has now.

3. Which of the following expressions are equal to 3 (2x - 4)

   a. 6x - 4
   b. 2 (3x -6)
   c. 2x - 12
   d. 5x - 7
   e. 6x - 12

4. There are 620 students in Kolman Middle School. If the number of boys is 160 more than the number of girls in the school, how many girls are in the school?

5. Dante wants to order some shorts from an online store. Shorts are $23.00 and shipping on all orders is $7.99. Dante has $80 to spend.

    a. Write an inequality to represent the situation.
    b. How many pairs of shorts can Dante order without exceeding his budget amount?

Need some hints? Look on page 176 for help

## Level 3 Algebra

1. Write the following as an algebraic equation:
    a. Twelve divided by a number minus three is one.

    b. Four times a number plus six is twenty-six.

2. Bethany says the following two expressions are equivalent.

    4b - 2 + 2b - 9 and 2b – 11

    Is she correct? Why or why not?

3. Katie and Ahmed work at a popcorn stand. Katie sells 23 bags per hour. Ahmed sells 26 bags per hour. This week, Ahmed sold an additional 16 bags. Write an expression to represent the weekly sales for both Katie (K) and Ahmed (A) this week.

    K =

    A =

4. Rico is an engineer for the city. His annual salary is $58,240. If Rico receives a paycheck every 2 weeks, what is his biweekly salary pay?

5. Find four consecutive integers whose sum is 78.

Need some hints? Look on page 177 for help

# Level 4 Algebra

1. Solve the following in scientific notation form

$$(6.4 \times 10^8) \div (1.6 \times 10^5)$$

2. An equation is shown

$$4^a \cdot 4^b = 4^{-1}$$

What are 2 possible values for both a and b?

3. The function y= 6.25x + 8 represents the total amount of money (y) Sara makes over (x) weeks. What is true about the function?

A. It is nonlinear because it increases at a constant rate.
B. It is nonlinear because it is always increasing.
C. It is linear because it is always increasing.
D. It is linear because it is increasing at a constant rate.

4. The graph shows how many lawns Seth mowed. The equation shows how many lawns Luca mowed. Who mowed more lawns?

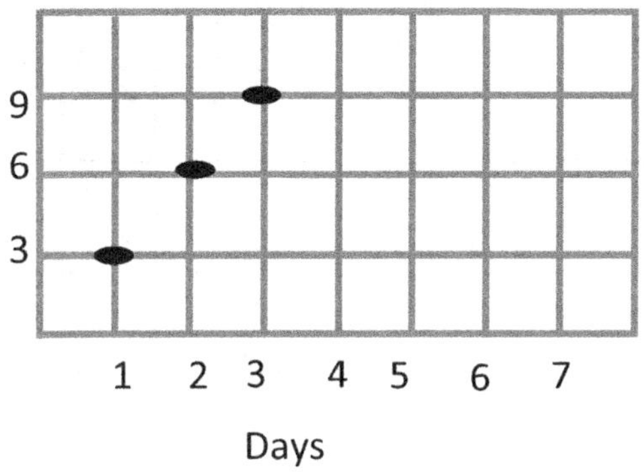


y = 4x

*x = days*

*y = lawns mowed*

5. Select if each equation has no solution, one solution, or many solutions.

| Equation | No solution | One solution | Infinite solutions |
|---|---|---|---|
| 9x = 9x – 1 | | | |
| 9x - 1 = 1 + 9x | | | |
| 9x - 1 = 9x – 1 | | | |

Need some hints? Look on page 177 for help

| SELF-ASSESSMENT ZONE<br>Level 1 | | | | | | |
|---|---|---|---|---|---|---|
| Question | Answer | Correct? | Lucky Guess | Knew It | Simple Mistake | Need Practice |
| 1 | 1 | ☐ | ☐ | ☐ | ☐ | ☐ |
| 2 | 3 (5+6) | ☐ | ☐ | ☐ | ☐ | ☐ |
| 3 | 9 and 13 | ☐ | ☐ | ☐ | ☐ | ☐ |
| 4 | 4 and 20: double or multiply by 2<br>30 and 30; add 7 | ☐ | ☐ | ☐ | ☐ | ☐ |
| 5 | $4.00 | ☐ | ☐ | ☐ | ☐ | ☐ |
| | Total | | | | | |
| What do you think is going well? | | | | | | |
| What do you need to do to improve? | | | | | | |

## SELF-ASSESSMENT ZONE
### Level 2

| Question | Answer | Correct? | Lucky Guess | Knew It | Simple Mistake | Need Practice |
|---|---|---|---|---|---|---|
| 1 | A. 3<br>B. 5<br>C. 1<br>D. 4<br>E. 2 | ☐ | ☐ | ☐ | ☐ | ☐ |
| 2 | 12 + s | ☐ | ☐ | ☐ | ☐ | ☐ |
| 3 | B and E | ☐ | ☐ | ☐ | ☐ | ☐ |
| 4 | 230 girls | ☐ | ☐ | ☐ | ☐ | ☐ |
| 5 | A. $23x + 7.99 \leq 80$<br>B. 3 | ☐ | ☐ | ☐ | ☐ | ☐ |
| | **Total** | | | | | |

What do you think is going well?

What do you need to do to improve?

## SELF-ASSESSMENT ZONE
Level 3

| Question | Answer | Correct? | Lucky Guess | Knew It | Simple Mistake | Need Practice |
|---|---|---|---|---|---|---|
| 1 | A. (12 / (x-3)) = 1<br>B. 4x +6 = 26 | □ | □ | □ | □ | □ |
| 2 | No, she is incorrect. She did not combine like terms, I should be 6b-11 | □ | □ | □ | □ | □ |
| 3 | K = 23h<br>A = 23h+16 | □ | □ | □ | □ | □ |
| 4 | $2,240 | □ | □ | □ | □ | □ |
| 5 | (n)+(n+1)+(n+2)+(n+3)=78<br>18, 19, 20, 21 | □ | □ | □ | □ | □ |
| | Total | | | | | |

What do you think is going well?

What do you need to do to improve?

| SELF-ASSESSMENT ZONE<br>Level 4 | | | | | | |
|---|---|---|---|---|---|---|
| Question | Answer | Correct? | Lucky Guess | Knew It | Simple Mistake | Need Practice |
| 1 | $4 * 10^3$ | □ | □ | □ | □ | □ |
| 2 | Any two integers added to get -1 (i.e. 3 and -4 or -1 and 0) | □ | □ | □ | □ | □ |
| 3 | D | □ | □ | □ | □ | □ |
| 4 | Luca | □ | □ | □ | □ | □ |
| 5 | No<br>No<br>Infinite | □ | □ | □ | □ | □ |
| | Total | | | | | |

What do you think is going well?

What do you need to do to improve?

# SET A GOAL

*Write down something you will do in the next week to become a stronger math student. Think about one thing you can commit to doing, something you can stick with.*

# Chapter 2

## Everyone DOES Achieve More Together

Learning a new skill or working to complete a task can feel lonely sometimes. But we are never really alone if we are willing to learn from the people around us.

Learning from others – and learning *with* others – is good way to find success. In fact, teamwork is one of the most important parts of a successful company, team or group.

In these stories, the authors share their own experiences about learning and working with others.

**Louis Gigliotti**
**Goodyear Employee Starting in 2016**

## Trust your team

*Louis learned very early in his career that asking for help – and helping others – is key to a successful team*

There will be times when you do not feel as knowledgeable or confident as others. You may feel intimidated or confused when starting something new. When that happens, remember that the people around you want you to be successful and are there to help.

*Working in teams is the only way to get things done if you truly want to be successful.*

As an intern at Goodyear, I would have never imagined how cordial and helpful my fellow associates were when it came to delivering my projects. I was able to ask all kinds of questions without being judged. People are there to help you and are willing to work with you to help get things done.

Working in teams is the only way to get things done if you *truly* want to be successful. It is important to remember that you cannot reach your goals without working with others. Therefore, remind yourself to always be willing to help others and offer your support when someone may be struggling with a task. You never know when you will need the favor returned! It's not about you, it's about the *team!*

**FUN FACT:** Louis got his start at Goodyear as an intern. He is not alone. Lots of people get their start as an intern. Hip-hop businessman, rapper and songwriter Sean "Puffy" Combs (aka P-Diddy) also got his start as an intern at Uptown Records while he was a college student.

# REFLECTION ZONE

*There will always be others who are new and who need help. How can you show others that you are willing to support them as a good teammate?*

**David Pellegrini**

**Goodyear Employee Starting in 1987**

## Have skills, will travel

*David's skills as a computer programmer and team player have taken him all over the world in ways he didn't expect*

When I started at Goodyear, little did I know where it would take me and who I would work with. I was hired as a computer programmer right out of college. I thought most of my time would be spent in an office at my desk. I was in for a pleasant surprise.

I've worked on teams with members from around the world. I have visited 20 countries and many places in the United States. It is quite challenging, yet rewarding, to work with teammates from all over the world. We get to solve problems with technology, and to see those solutions being used in factories, warehouses and offices.

*It is quite challenging. Yet rewarding, to work with teammates from all over the world.*

One of my favorite projects was installing software at a factory in Chile. Our team had people from Brazil, Chile, Columbia, Guatemala, Germany and the United States. We all worked hard together to start up the software and improve our plant. The team members continue to stay in touch, and some of us have worked together on other projects, too. Science, technology, engineering and math provide a rewarding and challenging path, and you never know where you may go and who you may work with. But one thing is certain: you will go farther if you go with a team.

**FUN FACT:** David may have worked in over 20 countries, but he has a long way to go if he wants to visit every country where The Goodyear Tire & Rubber Company operates. Goodyear actually does business in *more than 60 countries* around the world!

# REFLECTION ZONE

*If you could go anywhere in the world, where would you visit? What do you think you could learn from the people who live there?*

*Feeling inspired?*

# LEVEL UP

**TRY THESE PRACTICE PROBLEMS**

Each level is a little more challenging

Check your work at the end

Look for hints if you need them

## CHALLENGE SET 2: Geometry

### Level 1 Geometry

1. What is the measure of the missing angle? (Picture is not to scale)

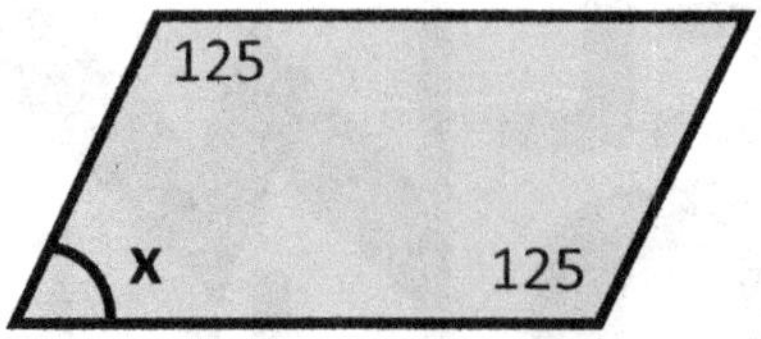


2. Select all the properties that a parallelogram and square share:

    a. At least 1 pair of parallel sides
    b. Have 4 sides
    c. Have 4 equal angles
    d. All angles add up to 360 degrees

3. Cubes were placed in a box. The volume of each cube is 1 cubic millimeter. What is the total volume of the box, in cubic millimeters?

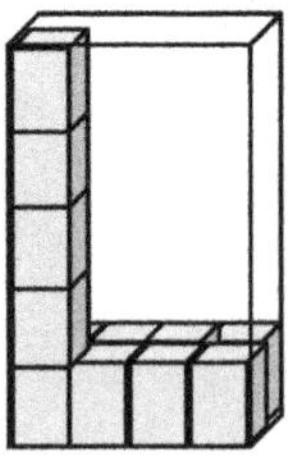

4. Max drove from his house to the grocery store. The location of the store is (5,2)

   - To drive home, he went 3 units left and 3 units up. Place the location of his home on the grid.

5. The volume of a rectangular prism is 48 $cm^3$. The height of the prism is 3. What are possible dimensions for the base of the prism?

Need some hints? Look on page 179 for help

# Level 2 Geometry

1. What is the distance between (-1,4) and (3,4) on the coordinate plane?

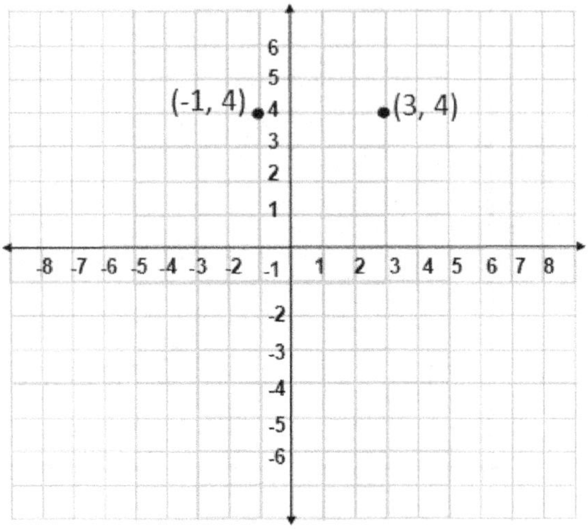


2. What is the surface area for the net of the rectangular prism?

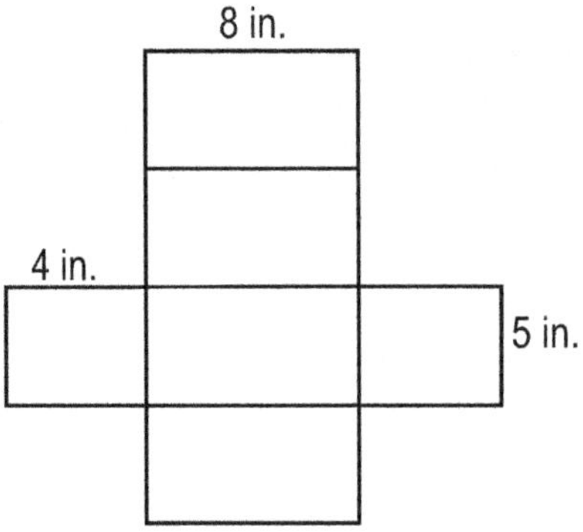

3. What is the volume of the figure?

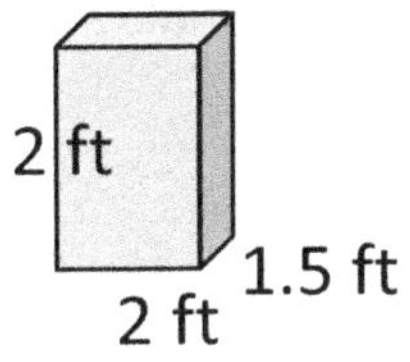


4. Plot the points for all given values on the number line.

$\frac{1}{4}$   -2   $-\frac{1}{2}$   $3\frac{3}{4}$

5. A small cube has a side length of 3 in. and a large cube has a side length of 7 in. What is the combined volume of the cubes?

Need some hints? Look on page 178 for help

## Level 3 Geometry

1. A triangle has a side length of 4 inches and a side length of 1 ½ inches. What is a possible length for the 3rd side of the triangle?

2. Draw the cross section of the triangular prism if the cut is perpendicular to the base.

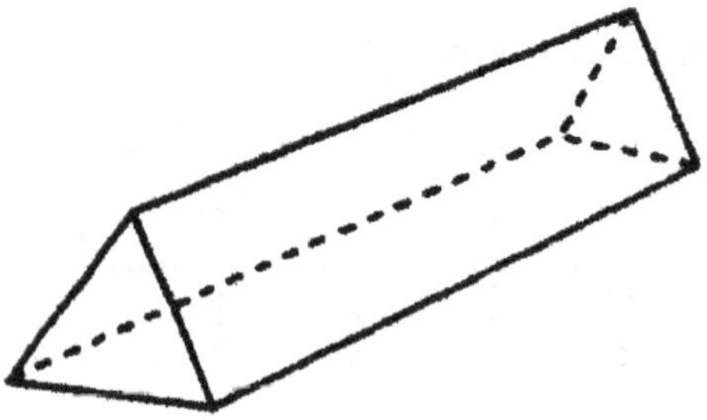

3. A circle has a diameter of 6.3 inches. What is the circumference of the circle, rounded to the nearest hundredth of an inch?

4. The following shows various angles coming from center point B.

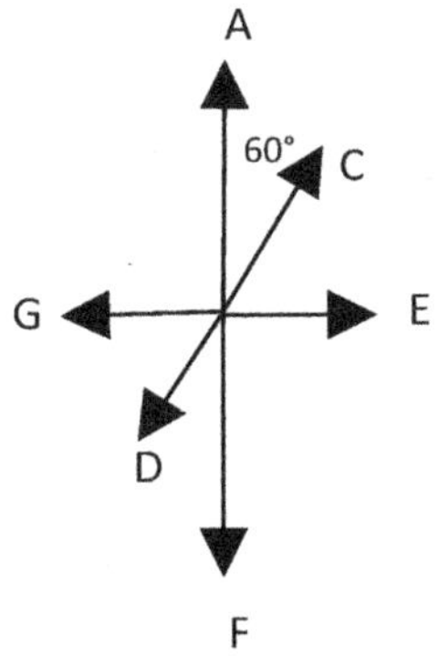


First label B; ABC= 60 degrees and is complementary to CBE. What is the measure of angle GBD?

5. The circumference of a circle is $15\Pi\ cm$. What is the area, in square centimeters, in terms of $\Pi$?

Need some hints? Look on page 179 for help

## Level 4 Geometry

1. How much greater is the volume in the first cylinder? Use 3.14 for $\pi$ and inches for units.

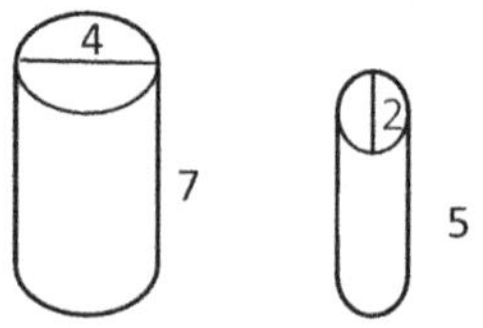


2. What is the value of y in terms of x if *a* is parallel to *b*?

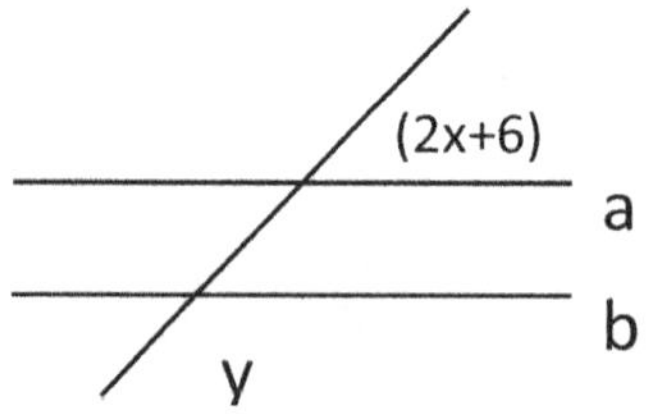


3. A right square pyramid has a height (h) of 6 inches and a side of the base (b) measuring 14 inches. What is the length of *a* in inches to the nearest tenth?

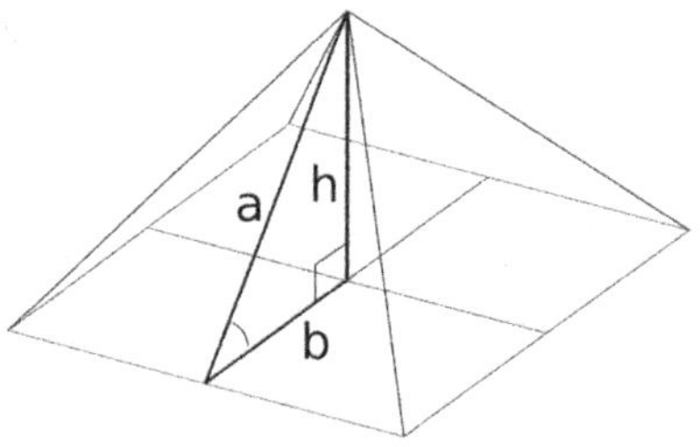

4. What is the value of x if the figure has been reflected and translated up?

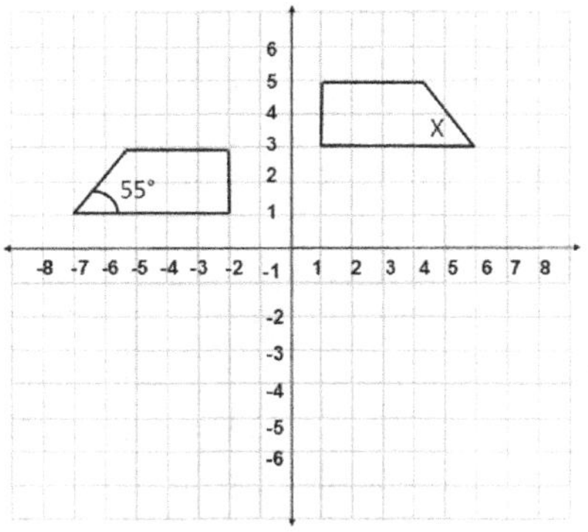


5. A triangle has side lengths of 2.2, 1.6, and 1.5 inches. What will the new lengths be if the triangle is dilated by 2 inches? Draw and label.

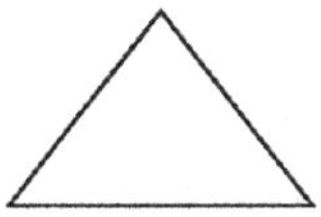

Need some hints? Look on page 179 for help

| SELF-ASSESSMENT ZONE<br>Level 1 | | | | | | |
|---|---|---|---|---|---|---|
| Question | Answer | Correct? | Lucky Guess | Knew It | Simple Mistake | Need Practice |
| 1 | 55° | ☐ | ☐ | ☐ | ☐ | ☐ |
| 2 | At least one pair of parallel sides, have for sides, and all angels add up to 360° | ☐ | ☐ | ☐ | ☐ | ☐ |
| 3 | 40 $\mathrm{mm}^3$ | ☐ | ☐ | ☐ | ☐ | ☐ |
| 4 | (2, 5) | ☐ | ☐ | ☐ | ☐ | ☐ |
| 5 | 4 cm and 4 cm; or<br>2 cm and 8 cm; or<br>1 cm and 16 cm | ☐ | ☐ | ☐ | ☐ | ☐ |
| | Total | | | | | |
| What do you think is going well? | | | | | | |
| What do you need to do to improve? | | | | | | |

## SELF-ASSESSMENT ZONE
### Level 2

| Question | Answer | Correct? | Lucky Guess | Knew It | Simple Mistake | Need Practice |
|---|---|---|---|---|---|---|
| 1 | 4 units | ☐ | ☐ | ☐ | ☐ | ☐ |
| 2 | 184 $in^2$ | ☐ | ☐ | ☐ | ☐ | ☐ |
| 3 | 6 $ft^3$ | ☐ | ☐ | ☐ | ☐ | ☐ |
| 4 | 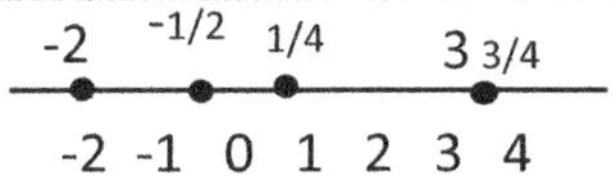  | ☐ | ☐ | ☐ | ☐ | ☐ |
| 5 | 200 $in^2$ | ☐ | ☐ | ☐ | ☐ | ☐ |
| | **Total** | | | | | |

What do you think is going well?

What do you need to do to improve?

## SELF-ASSESSMENT ZONE
### Level 3

| Question | Answer | Correct? | Lucky Guess | Knew It | Simple Mistake | Need Practice |
|---|---|---|---|---|---|---|
| 1 | Any rational number > 2 $^1/_2$ and < 5 $^1/_2$ | ☐ | ☐ | ☐ | ☐ | ☐ |
| 2 | | ☐ | ☐ | ☐ | ☐ | ☐ |
| 3 | 19.79 in | ☐ | ☐ | ☐ | ☐ | ☐ |
| 4 | 30° | ☐ | ☐ | ☐ | ☐ | ☐ |
| 5 | $56.25\pi\ cm^2$ | ☐ | ☐ | ☐ | ☐ | ☐ |
| | **Total** | | | | | |

What do you think is going well?

What do you need to do to improve?

## SELF-ASSESSMENT ZONE
### Level 4

| Question | Answer | Correct? | Lucky Guess | Knew It | Simple Mistake | Need Practice |
|---|---|---|---|---|---|---|
| 1 | 72.22 $in^2$ | □ | □ | □ | □ | □ |
| 2 | -2x + 174 | □ | □ | □ | □ | □ |
| 3 | 15.2 in | □ | □ | □ | □ | □ |
| 4 | 55° | □ | □ | □ | □ | □ |
| 5 | 4.4 in, 3.2 in, 3 in | □ | □ | □ | □ | □ |
| | Total | | | | | |

What do you think is going well?

What do you need to do to improve?

# SET A GOAL

*Think about the kind of person you want to be when you are older. Write down a vision for your best self. Then write down the kinds of habits you need to have NOW if you want to become that person. We gave you a little help.*

When I am older, I want to be the kind of person who...

I know I need to make these habits a part of my daily life:

# Chapter 3

## The Bank of YOU

Most people think of money when they hear the word "invest." But the best investment you can make in life is an investment in *yourself*.

These authors all understand that they need to invest in themselves if they want to grow and improve.

As you read these stories, think about how much you put in your own "bank account" of skills and experience.

**Sebastian Gruenheid**

**Goodyear Employee Starting in 2015**

## Be the author of your life

*Sebastian sees life as a book with blank pages. New skills and experiences fill each page*

Life is a book with white pages. You fill your book with your story. No matter what the story is, you are the author.

I am not originally from the United States. When I was a kid, I always wanted to see America. The challenge was that *I could not speak English*. I knew that if I wanted to be successful in America, I would need to learn the language.

*Learning a new language when you are an adult is very hard. But I was committed to writing my own life story.*

And so, I started learning English when I was 27. Learning a new langauge when you are an adult is very hard. But I was committed to writing my own life story. It took me *nine years*, but I finally met my goals.

I have been living in the United States since 2015. I work for a global company, and using English is a part of my daily life.

My book still has more white pages, but many have already filled with achievements and adventures. Each page I write in the book of my life gives me confidence for all future things to come.

> **FUN FACT:** Sebastian's "book of life" is nowhere near the longest book in the world. Madeleine and Georges de Scudery wrote Cyrus the Great about 500 years ago. Their story, published in 10 volumes, is over 13,000 pages long! Be glad you don't have to read that for homework.

# REFLECTION ZONE

*What is the next page you want to fill in your own life story? Write down something you see in your future that you want to make come true.*

**Kathryn Thompson**

**Goodyear Employee Starting in 2014**

# Learning without a teacher

*Kathryn had to "learn to learn" alone in her professional life*

A big part of working at my job is actually learning, just like in school. But instead of having a teacher guiding me to learn the right things at the right times, I am mostly in charge of my own learning process. That sounds awesome, right?

This can actually be very challenging. In school I could pay close attention to my lessons, study hard, and usually do well on my homework and tests. When I struggled or had to do something new, I had my teachers and my parents pushing me and challenging me to try, learn and improve. At work, I have to push myself when I'm assigned something I don't know how to do.

*I remind myself of how good it always feels when I overcome a challenge and learn something new.*

That may not sound like a big deal, but sometimes it can be scary to face a new challenge. I often find myself wanting to put off a new task because I worry that I won't be able to do it or I will mess up.

So what do I do? I remind myself of how good it always feels when I overcome a challenge and learn something new. And I remember how much better I will be as an engineer when I have this new skill.

When I overcome a challenge, I always look back and think that it wasn't as hard as I thought it would be. It is always scarier looking at a new challenge than it is actually facing it. The hardest part is making the decision to start, and before you know it, you will be saying, "Wow! That was easy!"

**FUN FACT:** Kathryn reminds herself how good it feels when she overcomes a challenge. She's right. Your brain actually releases chemicals that make you *feel happy* when you work through a challenge and meet with success. Don't give up – we promise you'll feel better.

# REFLECTION ZONE

*Write down one challenge you are facing on your own right now, even if it is a small challenge. Then write down how you will feel when you overcome that challenge.*

**Collin Bush**

**Goodyear Employee Starting in 2015**

## Become a professional "you"

*By focusing on his interests, Collin found the right career for him*

I grew up thinking that I would become a surgeon. I dreamed of performing surgeries to help athletes recover from their injuries. Once I started taking my college classes, I didn't feel that passion for becoming a surgeon like I used to. I was confused about what I wanted to pursue as a career. After all those years, it was hard to see myself as something other than a surgeon. How could I still help people, even if I wasn't in an operating room?

I decided to write down what I was passionate about. Soon, I realized that I love trying to solve complex problems. I am always thinking about how things can be improved. My passion is making bad things good, and good things even better.

*Once I defined my passion, I started looking for jobs where I could use that passion every day.*

Once I defined my passion, I started looking for jobs where I could use that passion every day. Working in Human Resources allows me to use my passion. I help our associates and our business teams work at their very best.

I love working for Goodyear because I get to solve complex problems and help people and the business do better. I love working for Goodyear because I get to follow my passion. A successful career isn't always determined by the product a company makes. A successful career is about finding a job and a company that lets you use your talents and follow your passion.

**FUN FACT:** Collin changed his mind about a career, from being a surgeon to working in human resources. That isn't uncommon. Most adults today will have 7 or 10 *different jobs* in their lives.

# REFLECTION ZONE

*Think about what you like to do when exploring careers. Do you like solving problems, writing, doing research? Working with others? Talking to lots of people, or keeping to yourself? Write down what you like to do and how you like to work.*

# Feeling inspired?

**TRY THESE PRACTICE PROBLEMS**

Each level is a little more challenging

Check your work at the end

Look for hints if you need them

## CHALLENGE SET 3: NUMBER SENSE

### Level 1 Number Sense

1. Find all equivalent fractions to ⅓

   3/9 4/8 4/3 5/15 2/6 30/100

2. Calvin wants to make banana muffins. The recipe calls for 2 cups of flour and 1 cup of sugar. He only has a quarter cup measurer. How many quarter cups of flour and quarter cups of sugar will Calvin need?

3. Write $8 \div 3$ as a fraction. Then, show this number on a number line

   < -----------------0-----1-----2-----3-----4-----5---------------- >

4. Alex is walking to Nick's house that is $\frac{7}{8}$ of a mile away. He has walked $\frac{1}{3}$ of a mile already. How much further does Alex need to walk to get to Nick's house?

5. Kendra bought $3\frac{4}{7}$ pounds of strawberries. What is the amount of strawberries Kendra bought rounded to the nearest $\frac{1}{2}$ lb?

Need some hints? Look on page 180 for help

## Level 2 Number Sense

1. What is the greatest common factor of 15, 24, and 48?

2. Jackie ordered a sub sandwich for her birthday party and cut the sandwich into 10 pieces. At the party, her friends ate 7 of the 10 pieces. Which model represents how many pieces of the sub were eaten?

7/100 70/100 3/100

3. Evaluate the expression $\frac{6}{0.4}$

4. Grant and Ben each bought their own small pizza. Grant ate $\frac{3}{8}$ of his pizza and Ben ate $\frac{1}{2}$ of his pizza. If they put the leftover pieces together, how much pizza do they have left?

5. One serving of cereal is $\frac{2}{3}$ of a cup. Sharla ate 2 cups of cereal. How many servings did she eat?

Need some hints? Look on page 180 for help

## Level 3 Number Sense

1. Simplify $\frac{2}{-3} + (\frac{-4}{9})$

2. Seth is dividing - 68 by a positive whole number. Which statement describes the quotient Seth should get?

   A. A positive greater than or equal to 68
   B. A negative greater than or equal to - 68
   C. A positive less than or equal to 68
   D. A negative less than or equal to - 68

3. Simplify - ½ + ($3^2$) - ( - 7.5)

4. Which expression is greater? Place both answers on the number line and label.

$.27 - \frac{32}{100}$ $\qquad$ $-.18 + \frac{68}{100}$

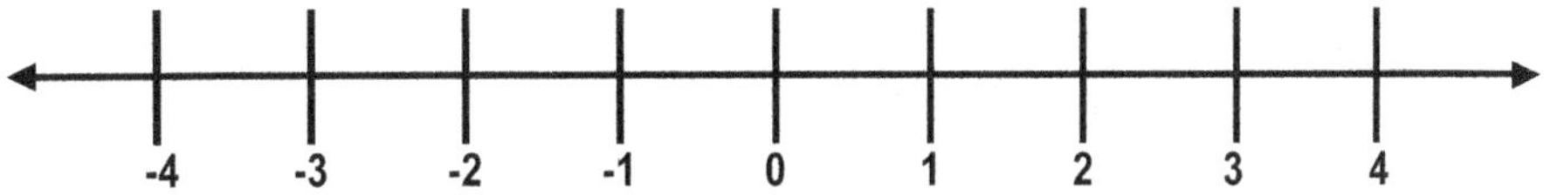


5. The average temperatures in Alaska for a week in February is 3 degrees. What are possible temperatures for Monday and Thursday?

| Day | Temp. |
|---|---|
| Sunday | -1 |
| Monday | |
| Tuesday | -7 |
| Wednesday | 3 |
| Thursday | |
| Friday | 2 |
| Saturday | 6 |

Need some hints? Look on page 181 for help

## Level 4 Number Sense

1. $^{2}/_{7}$ belongs to which subset of real numbers? Explain below why it does not fit the other subsets.

Rational     Natural     Integer     Irrational

______________________________________________

______________________________________________

______________________________________________

______________________________________________

______________________________________________

______________________________________________

2. A scientist can measure cells under a microscope. The bacteria cell measured .000017 mm. What is this in scientific notation?

3. What is the value of $\left(\frac{4+4^2}{5(2)} - 7\right) + \sqrt{121}$

4. Elana divides 10 by a number. The result was greater than 10. What number could she have used to divide 10 by? Explain how you know.

½ -5 2 ½ 1

5. Twenty-one out of twenty-eight students pack lunch for the field trip. What is the best estimate for the percent of students who packed?

30%

50%

70%

90%

Need some hints? Look on page 181 for help

| SELF-ASSESSMENT ZONE<br>Level 1 | | | | | | | |
|---|---|---|---|---|---|---|---|
| Question | Answer | Correct? | Lucky Guess | Knew It | Simple Mistake | Need Practice |
| 1 | 3/9, 5/15, 2/6 | ☐ | ☐ | ☐ | ☐ | ☐ |
| 2 | 8 quarter cups of flour and 4 quarter cups of sugar | ☐ | ☐ | ☐ | ☐ | ☐ |
| 3 | 8/3 the point should be put on the line between 2 and 3, more than halfway to 3 | ☐ | ☐ | ☐ | ☐ | ☐ |
| 4 | 13/24 of a mile | ☐ | ☐ | ☐ | ☐ | ☐ |
| 5 | $3\,{}^{1}/_{2}$ lbs. | ☐ | ☐ | ☐ | ☐ | ☐ |
| | Total | | | | | |

What do you think is going well?

What do you need to do to improve?

| SELF-ASSESSMENT ZONE<br>Level 2 | | | | | | |
|---|---|---|---|---|---|---|
| Question | Answer | Correct? | Lucky Guess | Knew It | Simple Mistake | Need Practice |
| 1 | 3 | ☐ | ☐ | ☐ | ☐ | ☐ |
| 2 | 70/100 | ☐ | ☐ | ☐ | ☐ | ☐ |
| 3 | 15 | ☐ | ☐ | ☐ | ☐ | ☐ |
| 4 | $1\,{}^{1}/_{8}$ pizzas | ☐ | ☐ | ☐ | ☐ | ☐ |
| 5 | 3 servings | ☐ | ☐ | ☐ | ☐ | ☐ |
| | Total | | | | | |
| What do you think is going well? | | | | | | |
| What do you need to do to improve? | | | | | | |

| SELF-ASSESSMENT ZONE<br>Level 3 | | | | | | | |
|---|---|---|---|---|---|---|---|
| Question | Answer | Correct? | Lucky Guess | Knew It | Simple Mistake | Need Practice |
| 1 | $-10/_9$ or $-1\ 1/_9$ | □ | □ | □ | □ | □ |
| 2 | A negative greater than or equal to -68 | □ | □ | □ | □ | □ |
| 3 | 16 | □ | □ | □ | □ | □ |
| 4 | .05 and .5<br>.5 is greater | □ | □ | □ | □ | □ |
| 5 | The two choices need to equal 3 when added together (i.e. 7 and -4, 6 and -3, 5 and -2) | □ | □ | □ | □ | □ |
| | Total | | | | | |

What do you think is going well?

What do you need to do to improve?

| SELF-ASSESSMENT ZONE<br>Level 4 | | | | | | |
|---|---|---|---|---|---|---|
| **Question** | **Answer** | **Correct?** | **Lucky Guess** | **Knew It** | **Simple Mistake** | **Need Practice** |
| **1** | Rational | □ | □ | □ | □ | □ |
| **2** | $1.7 * 10^{-5}$ | □ | □ | □ | □ | □ |
| **3** | 6 | □ | □ | □ | □ | □ |
| **4** | $^1/_2$ | □ | □ | □ | □ | □ |
| **5** | 70% | □ | □ | □ | □ | □ |
| | **Total** | | | | | |
| What do you think is going well? | | | | | | |
| What do you need to do to improve? | | | | | | |

# SET A GOAL

*Start with a penny. Double the value of what you have every day. If you do this for one week, you will have 64 cents. If you do this for one month, you will have over FIVE MILLION DOLLARS! This is the power of doubling.*

*Write down something you are working on right now – something you are trying to get better at. Write down how you could work twice as hard for the next week. Commit to it!*

# Chapter 4

## Every long journey starts with a single step

Many parts of life can feel like a long, impossible journey. There will be a time when you look at what's ahead and think, *There is no way I can do this!*

The authors in this story all felt that way at one time or another.

As you read these stories, look for lessons on how to stay positive and focused on your own journey.

**Barbara Madigan**

**Goodyear Employee Starting in 2002**

## Every step counts

*Barbra learned to take every challenge "one step at a time"*

The summer I was 16, my sister and I hiked the Bruce Trail in Ontario, Canada. It took five weeks. We hiked 430 miles, end-to-end. We had our tent and backpacks, a guidebook with maps, and a plan to hike 10 miles per day.

On the first day, we hiked *one mile*, and it *wore us out*. We spent over an hour setting up camp. The next morning, it took us two hours to make breakfast and pack up. I took a moment to look at the big map at the front of the guidebook and saw the 429 miles stretching out ahead of us. I saw the tiny one mile we had hiked the first day.

I almost gave up and went home.

But then I quickly folded up the big map and looked at our map for that day. I saw where we planned to eat lunch and where we planned to camp that night. I said, "We can do that!" And we did. For the next five weeks, each morning we planned that day's hike – where we'd stop for breaks and where we'd camp.

*Look at the whole picture so you can make a good plan.*

We had some difficult days. We got blisters, got lost, got mad at each other, and sometimes it was all we could do to make it to the next break, or even the next step. But we kept going, one day (or step) at a time, following the plan. And as we went, we got stronger, better at navigating, and we learned to set up camp in minutes.

I didn't look at the big map again until we had completed more than two-thirds of the trail. By then it was fun to see the progress we'd made, and I had the confidence we would finish it.

I learned a great lesson that day that has helped me through many tough times since. Look at the whole picture so you can make a good plan, but then just tackle it one day, or step, at a time.

**FUN FACT:** The longest hiking trail in the United States is the American Discovery Trail. The trail starts at the Atlantic Ocean in Delaware, and ends at the Pacific Ocean in California. The trail is 5,057 miles. To walk the whole trail, you would take about 10 million steps!

# REFLECTION ZONE

*Write down a goal for yourself that would take a long time to accomplish. Then write the "first step" you can take to reach that goal.*

**Kelly Reef**

**Goodyear Employee Starting in 2015**

## One potato, two potato – how many more?!?

*Kelly's got a life lesson from her grandfather that helps her still today. You couldn't "pick" a better story*

When I was 9, I was picking potatoes in our family garden with my grandfather. I remember whining and complaining about how long it was taking and how many potatoes there were left to pick.

My grandfather turned around to me and said, "Don't look ahead at all those potatoes that need picking, *turn around* and look at all those potatoes you already picked."

*Turn around and look at all those potatoes you already picked.*

From then on, when I was working on homework or a project that seemed hard or too time consuming, I stopped and looked back at everything I had already learned. This has given me the confidence to pick that next potato from the garden or to tackle that next math problem, one step at a time.

**FUN FACT:** The average American eats about 124 pounds of potatoes every year. The average person living in Germany eats about 250 pounds of potatoes every year!

# REFLECTION ZONE

*What is one "potato" in your life? Write down a job you might avoid doing because it's hard. Then write down how you will finish it.*

**Kristen Delpinal**

**Goodyear employee Starting in 2013**

## "Stick" with it on large tasks

*Big challenges don't overwhelm Kristen now that she has a strategy*

Being faced with a complex project with a fast-approaching deadline can be overwhelming and stressful for anyone. When there's a large challenge ahead, it can be difficult to know where to start or to know when to stop. Over the years, I've developed a strategy that helps me get organized and focus my energy and limited time in the most productive way.

*When you break a large challenge into small tasks, it becomes more manageable and achievable.*

First, I break the challenge into smaller tasks and write each one on its own sticky note. Then, I prioritize the tasks in order on the wall, starting with the quickest and easiest tasks whenever possible. Once they're all in order, I set a timer for my first task and get started!

When you break a large challenge into small tasks, it becomes more manageable and achievable. By establishing expectations for yourself, you give each part of a challenge the time and attention it deserves. In the end, you'll be able to complete your work and feel satisfied that you faced your challenge head on and gave it your best.

**FUN FACT:** Sticky notes were invented by the 3M Company. They were made from glue that was too weak for anything else. That weak glue became a huge success. 3M sells over 50 billion Post-it® notes each year!

# REFLECTION ZONE

*How do you "break it down" when facing a big task? Do you use a calendar, notes, smart devices or something else? What is your plan to stay on track and focused through challenges?*

*Feeling inspired?*

# LEVEL UP

**TRY THESE PRACTICE PROBLEMS**

Each level is a little more challenging

Check your work at the end

Look for hints if you need them

## CHALLENGE SET 4: DATA AND STATISTICS

### Level 1 Data and Statistics

1. What type of graph would be the best choice to display Anna's height from 6th grade to 7th grade?

    A. bar graph
    B. stem and leaf
    C. line graph
    D. line plot

2. The table lists the weights of the wrestlers for a tournament. How many wrestlers are in the tournament?

| weight | 80 -84 | 85 -90 | 91 – 95 | 96 - 100 |
|---|---|---|---|---|
| wrestlers | 11 | 16 | 23 | 17 |

3. Brent caught the following number of fish each day on a trip for a week:

    3, 1, 0, 4, 1, 2, 1

    Use the data to complete the line plot:

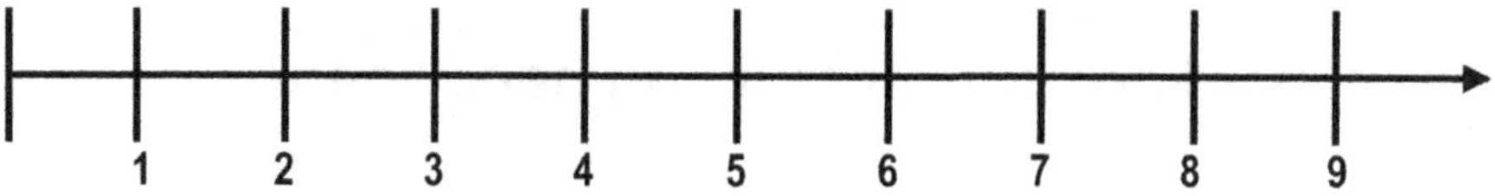

4. These numbers are values in a data set: 3, 6, 1, 8, 3, 6, 5. The number 8 represents what fraction of the data?

5. Give an example scenario for the double bar graph provided.

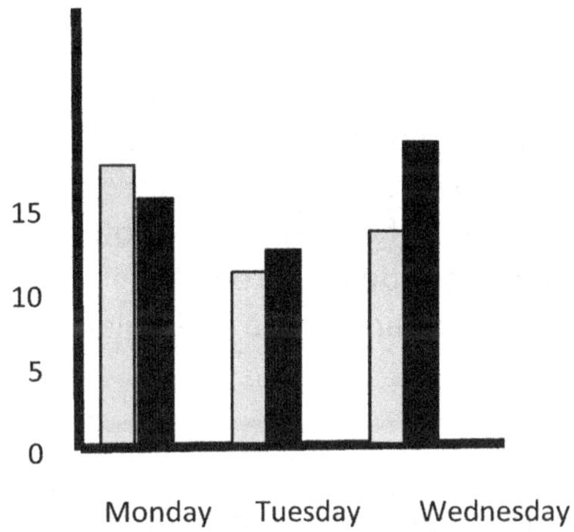


Gray: ___________

Black:___________

_______________________________________________

_______________________________________________

_______________________________________________

_______________________________________________

Need some hints? Look on page 182 for help

# Level 2 Data and Statistics

1. How many more squares and circles are needed to make the ratio of circles to total be 9:15?

2. Ally wants to ask her team a statistical question. Select all questions that are statistical.

   A. What day do you prefer to practice?
   B. Who is your coach?
   C. How many years of experience do you have?
   D. What color jersey do you prefer?
   E. How many games are on the schedule?

3. What is the mean, median, and mode for the set of data on the height of students in Mr. Calloy's class?

| **Student** | **Height** |
|---|---|
| Alexis | 42 " |
| Bradley | 38" |
| Hunter | 40 " |
| Scarlett | 40 " |
| Kiara | 34 " |

4. 23 students took a math test. Create a histogram to display the data:

85, 72, 76, 81, 64, 97, 95, 85, 86, 80, 77, 68,

93, 91,97, 82, 70, 76, 79, 85, 88, 91, 73

5. The stem and leaf plot shows the amount of minutes teens played on a video game system in 1 day.

| | |
|---|---|
| 3 | 1 4 |
| 4 | 4 5 8 9 |
| 5 | 2 3 5 8 |
| 6 | 1 2 6 |
| 7 | 4 |
| 8 | 3 5 7 |

A. How many teens reported minutes to the survey?
B. How many teens played less than 55 minutes?
C. What's the mean for the number of minutes played (rounded to the nearest whole number)?

Need some hints? Look on page 182 for help

## Level 3 Data and Statistics

1. A school wants to know what food students prefer for an everyday lunch choice, pizza or subs? Which survey would produce the best random sample?

    A. Survey all students who attend the Friday football game
    B. Survey every 5th person at the local supermarket
    C. Survey all students in STEM club
    D. Survey every tenth student entering the school on Monday morning

2. Estimate the number of times you need to roll a dice to get the number two as the outcome 10 times. Explain your thinking.

3. A bubble gum machine was just filled with 100 gumballs. There are 20 green, 10 pink, 10 white, 30 yellow, and 30 red. What is the probability the first gumball out will be green?

4. Which event is most likely to occur?

- ❑ Spinning a spinner divided into 3 sections labeled 1-3 and landing on a number greater than 1
- ❑ Flipping a coin with sides of heads or tails, and landing on heads
- ❑ Rolling a number cube with faces labeled 1-6, and the cube landing on an even number
- ❑ Pulling the eight of clubs card randomly from a standard deck of 52 cards

5. The results of a survey of 80 students selected randomly to share how they get to school are listed below:

- 24 students are car riders
- 41 students ride the bus
- 15 students walk

The total student body is 250. Based on the data, what is the best approximation for the total number of students who are car riders? Justify your answer.

Need some hints? Look on page 183 for help

# Level 4 Data and Statistics

1. Which situation best describes the information displayed on the scatterplot?

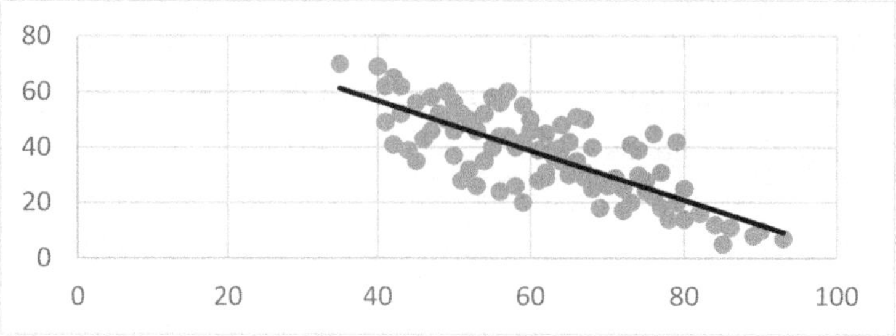


A. Positive correlation
B. Negative correlation
C. No correlation
D. Positive nonlinear correlation

2. Select all data sets which include an outlier

A. [2,3,6,2,5,4,6,3,4]
B. [6,7,6,1,4,6,7,4,5,6]
C. [3,4,7,9,3,4,7,9,3,4,7,9]
D. [1,1,2,1,1,1,0,1,2,0]
E. [8,10,16,11,7,6,8,12,10,7]

3. The test scores of 7 students are 85, 81, 98, 74, 67, 83, and 85. What is the range of the points scored by the 7 students?

4. Max has 5 coins in his pocket, 4 dimes and 1 nickel. He pulls a coin at random, looks at it, and puts it back. He selects another coin at random from the same pocket. What is the probability Max selected a dime both times?

5. The circle graph shows the percent of flowers sold at the florist for the month of April. The number of roses sold was 120. What is the number of tulips sold?

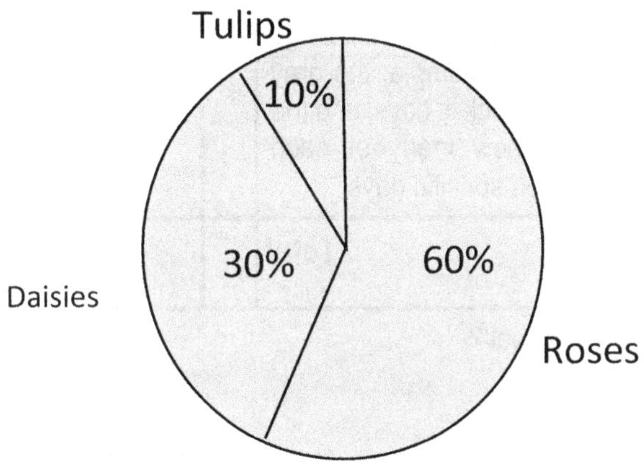


Need some hints? Look on page 183 for help

## SELF-ASSESSMENT ZONE

**Level 1**

| Question | Answer | Correct? | Lucky Guess | Knew It | Simple Mistake | Need Practice |
|---|---|---|---|---|---|---|
| 1 | Line graph | ☐ | ☐ | ☐ | ☐ | ☐ |
| 2 | 67 wrestlers | ☐ | ☐ | ☐ | ☐ | ☐ |
| 3 |  | ☐ | ☐ | ☐ | ☐ | ☐ |
| 4 | $^1/_7$ | ☐ | ☐ | ☐ | ☐ | ☐ |
| 5 | Answers can vary. Gray and black must represent comparing two things and the comparison should relate to a number that can change daily. Here is an example: the gray is girls and the black is boys, and the graph shows how many of each bought lunch on specific days. | ☐ | ☐ | ☐ | ☐ | ☐ |
| | **Total** | | | | | |

What do you think is going well?

What do you need to do to improve?

## SELF-ASSESSMENT ZONE
### Level 2

| Question | Answer | Correct? | Lucky Guess | Knew It | Simple Mistake | Need Practice |
|---|---|---|---|---|---|---|
| 1 | 3 circles and 2 squares | ☐ | ☐ | ☐ | ☐ | ☐ |
| 2 | A, C, and D | ☐ | ☐ | ☐ | ☐ | ☐ |
| 3 | Mean 38.8, Median 40, Mode 40 | ☐ | ☐ | ☐ | ☐ | ☐ |
| 4 | 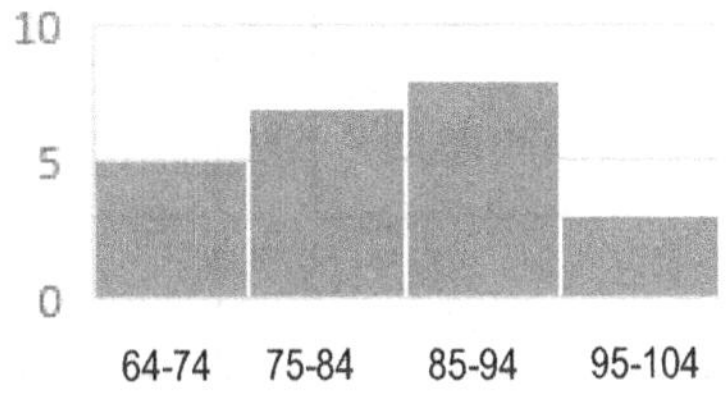  | ☐ | ☐ | ☐ | ☐ | ☐ |
| 5 | A. 17<br>B. 8<br>C. 58 | ☐ | ☐ | ☐ | ☐ | ☐ |
| **Total** | | | | | | |

What do you think is going well?

What do you need to do to improve?

| SELF-ASSESSMENT ZONE<br>Level 3 | | | | | | |
|---|---|---|---|---|---|---|
| **Question** | **Answer** | **Correct?** | **Lucky Guess** | **Knew It** | **Simple Mistake** | **Need Practice** |
| 1 | D | □ | □ | □ | □ | □ |
| 2 | 60 times, the probability is ⅙ so the equivalent would be 10/60 | □ | □ | □ | □ | □ |
| 3 | $^1/_5$ | □ | □ | □ | □ | □ |
| 4 | Spinning the spinner | □ | □ | □ | □ | □ |
| 5 | Approximately 75 students | □ | □ | □ | □ | □ |
| | **Total** | | | | | |
| What do you think is going well? | | | | | | |
| What do you need to do to improve? | | | | | | |

| SELF-ASSESSMENT ZONE<br>Level 4 | | | | | | |
|---|---|---|---|---|---|---|
| **Question** | **Answer** | **Correct?** | **Lucky Guess** | **Knew It** | **Simple Mistake** | **Need Practice** |
| 1 | B | ☐ | ☐ | ☐ | ☐ | ☐ |
| 2 | B and E | ☐ | ☐ | ☐ | ☐ | ☐ |
| 3 | 67 – 98 which is 31 | ☐ | ☐ | ☐ | ☐ | ☐ |
| 4 | $^{16}/_{25}$ | ☐ | ☐ | ☐ | ☐ | ☐ |
| 5 | 20 tulips | ☐ | ☐ | ☐ | ☐ | ☐ |
| | **Total** | | | | | |
| What do you think is going well? | | | | | | |
| What do you need to do to improve? | | | | | | |

# SET A GOAL

*Mistakes and failure are a part of the learning process. Mistakes are a part of life! As long as we learn from our mistakes, we can turn a setback into a great opportunity.*

*Write down a goal for how you will think differently or react differently the next time you have a setback. What will you do in this next week when things do not go your way?*

# Chapter 5

## Hard work beats talent

There is an old saying: *Hard work beats talent when talent doesn't work hard.*

The authors in these stories learned the value of hard work early in life, and that has paid off for them as adults.

Read these stories and think about how hard work now will make you stronger in life later.

**Trevor Moody**

**Goodyear Employee Starting in 2017**

## Goals + Hard Work = Success

*Having a goal is important, but Trevor knew he also needed to take action if he wanted to succeed*

When I walked into middle school for the first time, I set goals for myself to achieve excellence in the classroom, especially in mathematics. I knew from my experience in the sport of wrestling that setting goals would not be enough for success. Attaining my goals would require hard work in the form of extra study hours and asking questions when necessary.

During my seventh grade year in middle school, I enrolled in Pre-Algebra. I knew it would be a challenge, so I formed a game plan on how to be successful. Every morning 30 minutes before school started, I would meet with my Pre-Algebra teacher to review my homework from the previous day. This allowed me to understand the material to a greater degree, as I had my questions and doubts answered before a test or quiz.

The lessons of going the extra mile in the classroom and on the wrestling mat presented opportunities that I never thought were possible. I earned the chance to wrestle for Indiana University during college, and now I work for a tremendous company: Goodyear.

*Setting goals would not be enough for success. Attaining my goals would require hard work.*

Goals and aspirations are great, but they need to be connected with hard work if you want to be successful.

**FUN FACT:** Wrestling is considered the oldest sport in the world. You can find cave drawings of wrestlers that are over 5,000 years old. In ancient mythology, Zeus became the ruler of Earth by winning a wrestling match.

# REFLECTION ZONE

*Goals are only one part of success. Putting in effort is an important habit as well. Write down how you will "go the extra mile" in the next week.*

**Kathy DeRita**

**Goodyear Employee Starting in 2007**

## Extra credit? Yes, please!

*Aside from improving her grades, Kathy learned that extra work helped her improve herself*

When I was a kid, sometimes teachers would offer extra credit assignments. Extra credit can help kids who need to improve their grades. I always did the extra credit, even if I already had a good grade. I just never knew when I might need a few extra points.

All that extra work paid off. Sometimes it prevented me from going down a grade, and sometimes it bumped me up to the next grade. I always learned something new from doing extra work, which was a bonus!

As an adult I still try to do a little extra where I can, not because I'll get any extra points for it, but because it's a good thing to do. "Extra" might be in helping someone, or making a donation, or just sharing a simple smile and a kind word.

*I always learned something new from doing extra work, which was a bonus!*

Always do a little more than expected. You never know when it might make a difference.

**FUN FACT:** Kathy still tries to do something "extra" in life – like making a donation. She isn't alone. Americans are some of the most generous people in the world, giving about $360 billion each year to charity.

# REFLECTION ZONE

*Think about how you will do something "extra" for yourself or someone else. Write down one way you can help yourself or others by doing more than is expected.*

**Stephanie Wilton**

**Goodyear Employee Starting in 2012**

## More than minimum wage

*Stephanie didn't always love the jobs she had in high school, but those simple jobs became valuable experiences*

When I was in high school and college, I did a couple jobs here and there to earn some spending money. I would babysit in my neighborhood on Friday and Saturday nights. Then, I got a job working in an ice cream store.

I used to think those jobs were an inconvenience, mostly in the way of my social life and time to relax. I would roll my eyes at my parents when we would start arguing about me not wanting to go to work.

*Looking back, I now realize how priceless those first jobs were, and how they shaped me into who I am today.*

Looking back, I now realize how priceless those first jobs were, and how they shaped me into who I am today. I learned invaluable skills that I still use today, like learning how to work with people and how to manage my time.

My personal belief is that your experiences in life are what you make them. This is still a value that I try to hold true today to keep me focused on various tasks and projects that I am presented with at work. Don't squander and reject opportunities that seem dumb and pointless. You may just end up learning a thing or two, like I have been able to do.

**FUN FACT:** Fewer high school students have jobs than in the past. About 30 years ago, more than half of all teens had a part-time or summer job. Now only about 30% of teens have a job.

# REFLECTION ZONE

*List one skill or habit you are learning now that will make you a better person later in life.*

*Feeling inspired?*

# LEVEL UP

**TRY THESE PRACTICE PROBLEMS**

Each level is a little more challenging

Check your work at the end

Look for hints if you need them

## CHALLENGE SET 5: ALGEBRA

### Level 1 Algebra

1. Evaluate the expression $5 + (12 \div 4) - 7$

2. What is the numerical expression for *The quotient of twenty and eight less than twelve?*

3. Given the following information, complete the chart:

   Sequence a: Start with 5 and add 1

   Sequence b: Start with 11 and add 6

| A | B |
|---|---|
| | |
| | |
| | |
| | |

4. Find the missing numbers to complete the math sentences.

$24 = 6 + \_\_\_\_ \times 2$ $\qquad$ $\_\_\_\_\_\_ - 20 \div 5 = 12$

5. Which problem could be solved with the expression
   $2 \times (3 + 2 + 2)$ ?

   A. David was collecting flowers for his mom. He picked 3 roses and 2 daisies. He then decided he would pick another set of the same flowers for his sister. How many flowers did he pick?
   B. Olivia needed to practice her instrument for the concert. For 5 weeks she practiced for 3 hours on Monday, 2 hours on Wednesday, and 2 hours on Friday. How many hours did she practice in all?
   C. Gavin wanted to earn money doing chores. He earned 3 dollars for cleaning his room, 2 dollars for taking out the trash, and 2 dollars for feeding the family dog. He did the same chores for 2 weeks. How much did Gavin earn?

Need some hints? Look on page 184 for help

# Level 2 Algebra

1. Solve

   $(2^2 + 4^2) \times 2 - 5^2$

2. The tightrope walker must practice a minimum of 12 hours on the 3 foot line before moving up to the 6 foot line. Create an inequality to represent the situation.

3. Which of the following expressions are equal to $8y$- 12x + 24

   a. 3(8y - 4x + 8)
   b. 2(4$y$- 6x + 12)
   c. 2(4y - 6 + 24)
   d. 8(y - 4x + 3)
   e. 4(2y - 3x + 6)

4. There are 208 books for the students to choose from. If the number of fiction books is 84 more than the number of non fiction books in the library, how many fiction books are in the library?

5. Marco earned $8.75 per hour working with the landscaping company for the summer. Write an equation to find out how much total money he makes (m) for a set amount of hours (h). Then find out how many hours Marco would have to work to make $200. Round your answer to the nearest whole number.

Need some hints? Look on page 184 for help

# Level 3 Algebra

1. What is the simplified expression for the product of 4 and ($\frac{1}{4}$ x + 8)?

2. Which of the following are equivalent expressions to -2(5x - 3) + 7x ?

    a. -3x +6
    b. 10x - 6 + 7x
    c. -10x - 6 + 7x
    d. -10x + 6 + 7x
    e. 3x - 5 + 7x
    f. 10x - 5

3. Tyler was going bike riding on vacation. Instead of bringing his own bike, he decided to rent one at the hotel. The cost was $10, plus $2.50 for every hour he had the bike. Tyler paid a total of $18.75 for his rental. How long did he rent the bike?

2. Amari and Olivia were shopping at the mall. They each bought the same shirt. The shirt was $26.95 but the tag said 15% off. Which expression(s) show the price of the shirt each girl paid with the discount?

- ❑ 26.95 - .15
- ❑ .85(26.95)
- ❑ 26.95 - 26.95(.15)
- ❑ 26.95 - 26.95(1.5)

3. Find three consecutive integers whose sum is -24.

Need some hints? Look on page 185 for help

## Level 4 Algebra

1. Solve the following in scientific notation form

$$(8.7 \text{ x } 10^3)\ x\ (\ 3.2 \text{ x } 10^2)$$

2. An equation is shown:

$$3^a \cdot 3^b = \frac{1}{3^3}$$

What are 2 possible values for both a and b?

3. What is the value of x in the equation 9(x - 6) - 6(½ x + 3) = 14?

4. Which table(s) represent(s) a function?

A.

| X | y |
|---|---|
| 3 | 5 |
| 4 | 7 |
| 5 | 9 |
| 3 | 11 |

B.

| X | y |
|---|---|
| 1 | 4 |
| 2 | 4 |
| 3 | 4 |
| 4 | 4 |

C.

| X | y |
|---|---|
| 7 | 8 |
| 9 | 10 |
| 12 | 13 |
| 13 | 14 |

5. The theatre displays a table to show how much money was in the box after selling tickets. If all tickets cost the same price, write an equation to represent the situation and use the equation to find out how much money they started with in the collection box.

| Number of tickets sold | Amount of $ in the box |
|---|---|
| 12 | $167.25 |
| 21 | $273 |
| 36 | $449.25 |

Need some hints? Look on page 185 for help

| SELF-ASSESSMENT ZONE<br>Level 1 | | | | | | | |
|---|---|---|---|---|---|---|---|
| Question | Answer | Correct? | Lucky Guess | Knew It | Simple Mistake | Need Practice |
| 1 | 1 | □ | □ | □ | □ | □ |
| 2 | $\frac{20}{(12-8)}$ | □ | □ | □ | □ | □ |
| 3 | A: 5, 6, 7, 8<br>B: 11, 17, 23, 29 | □ | □ | □ | □ | □ |
| 4 | 9, 16 | □ | □ | □ | □ | □ |
| 5 | C | □ | □ | □ | □ | □ |
| | Total | | | | | |

What do you think is going well?

What do you need to do to improve?

## SELF-ASSESSMENT ZONE
### Level 2

| Question | Answer | Correct? | Lucky Guess | Knew It | Simple Mistake | Need Practice |
|---|---|---|---|---|---|---|
| 1 | 15 | ☐ | ☐ | ☐ | ☐ | ☐ |
| 2 | $X \geq 12$ | ☐ | ☐ | ☐ | ☐ | ☐ |
| 3 | B and E | ☐ | ☐ | ☐ | ☐ | ☐ |
| 4 | 146 fiction books | ☐ | ☐ | ☐ | ☐ | ☐ |
| 5 | 23 hours | ☐ | ☐ | ☐ | ☐ | ☐ |
| | **Total** | | | | | |

What do you think is going well?

What do you need to do to improve?

## SELF-ASSESSMENT ZONE
### Level 3

| Question | Answer | Correct? | Lucky Guess | Knew It | Simple Mistake | Need Practice |
|---|---|---|---|---|---|---|
| 1 | X + 32 | ☐ | ☐ | ☐ | ☐ | ☐ |
| 2 | A and D | ☐ | ☐ | ☐ | ☐ | ☐ |
| 3 | 3.5 Hours | ☐ | ☐ | ☐ | ☐ | ☐ |
| 4 | .85(26.95) and 26.95-26.95(.15) | ☐ | ☐ | ☐ | ☐ | ☐ |
| 5 | -9, -8, -7 | ☐ | ☐ | ☐ | ☐ | ☐ |
| | Total | | | | | |

What do you think is going well?

What do you need to do to improve?

## SELF-ASSESSMENT ZONE
### Level 4

| Question | Answer | Correct? | Lucky Guess | Knew It | Simple Mistake | Need Practice |
|---|---|---|---|---|---|---|
| 1 | $2.784 * 10^6$ | □ | □ | □ | □ | □ |
| 2 | Any two integers added equal -3 (i.e. 5 + (-8); (-1)+(-2)) | □ | □ | □ | □ | □ |
| 3 | $\frac{43}{3}$ | □ | □ | □ | □ | □ |
| 4 | B and C | □ | □ | □ | □ | □ |
| 5 | $26.25 | □ | □ | □ | □ | □ |
| | **Total** | | | | | |

What do you think is going well?

What do you need to do to improve?

# SET A GOAL

*You are now more than halfway through this book. Set a goal for how you will do on the Practice Two Test. Write down the kind of score you want to get, and how you will work to get that score.*

# Chapter 6

## A place for everything – and everyone

When you think about it, feeling out of place is one thing you have in common with *everyone*.

These stories are about how the authors felt out of place in their own lives, and what they did to overcome those feelings.

As you read, think about times when you have felt out of place. Also, give some thought to how you could help others feel like they belong.

**Jamie Dunphy**

**Goodyear Employee Starting in 1989**

## Welcome to the team – maybe

*What if you were the first person in a group who was different from everyone else? You can still fit in*

Fitting in can be difficult... especially when you are different from others on your team.

When I was in high school, we had an all-boys tennis team. There weren't enough players to compete against other schools, so one of my teachers asked me if I would be interested in joining the team. I loved tennis, so I told my teacher that I would be happy to join.

*I decided that if wanted to "fit in," I would have to approach my teammates myself*

I can remember the look on my teammates' faces on the first day of practice. No one really wanted to practice with a girl, but my coach made me feel welcome. I decided that if wanted to "fit in," I would have to approach my teammates myself. I began talking to them, complimenting them on their key strengths and asking them for help to improve my own game. They didn't say no.

Once the other players got to know me and understood that I had a passion for the same game as them, we all became friends. During my second year, three other girls joined our team. I had the courage to take the first step, and others followed. We didn't call it the "boys tennis team" after that. It was just our "high school tennis team." We all became very close friends, and to this day, I stay in touch with my former teammates.

Being the first of anything can be challenging. Be yourself, work hard and reach out to the people around you. You can be the one who makes the world better for those who follow you.

**FUN FACT:** Varsity tennis ranks number 7 in the top 10 most popular sports among high school girls, with 188,000 participants. Track is the most popular sport for girls, with almost half a million girls running, jumping or throwing every spring.

# REFLECTION ZONE

*Write down one thing you can remember the next time you feel out of place – something to remind yourself that you* ***do belong****.*

**Jake Jarrell**

**Goodyear Employee Starting in 2015**

## Who do you want to be? Start with being yourself

*Jake has learned that the key to making friends is "be yourself"*

I've used a variety of strategies to fit in throughout different stages in my life: middle school, high school, college, and even now at Goodyear. Some of these strategies include finding similar interests and using humor to break the ice. There are, however, a couple of strategies that have always provided the best results for me, regardless of the time, place, or people – *be genuine* and *be yourself*.

If you just be yourself, others will see your best qualities that make you unique. Then, you can connect with others on a more personal level and form the types friendships that will make your life more enjoyable.

I still talk to friends from every stage of my life. These lasting relationships have helped me when I had a challenging exam in school or when something stressful was happening at work. I can fall back on my friends and know they will be there to help through good times and bad.

> *If you just be yourself, others will see your best qualities that make you unique.*

Now that I've been at Goodyear for nearly three years, I've formed friendships that I know will last for years to come. These friendships didn't result from me giving a great presentation or coming up with the next great idea. These friendships happened just because I asked someone how their day was or I made a silly joke that someone laughed at. Be yourself and show you genuinely care about others. You'll be amazed by how many great people become a part of your life.

**FUN FACT:** "Break the ice" means to make people comfortable when they first meet. A ship that gets stuck in frozen waters needs help to "break the ice" and start moving again. The largest ice-breaking ship in the world is 524 feet long!

# REFLECTION ZONE

*You have something that makes you special. Everyone does. Write down something about your personality that you like. What do you want people to know about the kind of person you are?*

**Debbie Mathie**

**Goodyear Employee Starting in 1987**

## Experience: the best teacher

*Debbie didn't want to be an expert at making friends, but her life experiences can help you take the first steps with new people*

As I grew up, my mom, a single parent, worked very hard to provide a home for us in a safe neighborhood and decent apartment. But that often meant moving every 12 months due to changing leases or changing neighbors. And changing apartments also meant changing schools and making new friends.

It can be a very scary thing to make new friends every single school year. I had to figure out how to make friends – fast! The one thing I found when making new friends is a smile can make the first meeting easier. I found the other person is often just as scared as you, and greeting them with a smile signals the space is safe and welcoming.

I also found that paying attention to the other person you are with will give you clues to start a conversation. You might like their shoes (I always notice shoes!) or what they are having for lunch. Many times, complimenting the other person will make him or her feel good and continue the conversation. When that happens, keep asking questions about hobbies, family and pets. This is often where you find your common interests, and then, a friendship is born.

*The one thing I found when making new friends is a smile can make the first meeting easier.*

Another way to fit in easier is to be funny. This doesn't mean making a joke at someone else's expense; instead, share a funny situation that everyone can laugh about. Maybe it's a story about how you were rushing out in the morning and now you are wearing two different black shoes! (This is a real story and just happened to me!) I had more laughs with people I didn't know that day just laughing about my two different shoes!

Fitting in is about making other people feel comfortable around you. They don't know you, but if you can smile, laugh and learn something about them, they will welcome you into their world!

> **FUN FACT:** Debbie seems very interested in shoes. She isn't alone. A study in 2013 found that the average woman over the age of 18 owns 20 pairs of shoes!

# REFLECTION ZONE

*Write down something you have in common with a friend. How did this help you two connect?*

# Feeling inspired?

# LEVEL UP

**TRY THESE PRACTICE PROBLEMS**

Each level is a little more challenging

Check your work at the end

Look for hints if you need them

## CHALLENGE SET 6: GEOMETRY

### Level 1 Geometry

1. Select all the properties that a rectangle and trapezoid share:
    a. At least 1 pair of parallel sides
    b. Have 4 sides
    c. Have 4 equal angles
    d. All angles add up to 360 degrees

2. What is the measure of the missing angle (X)?

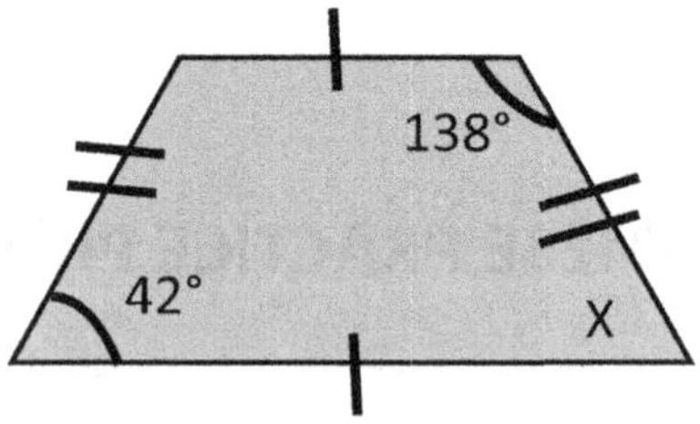


3. Tyler walked to the playground to meet his friends. The location of the playground is (1, 3). To drive home, he went 3 units right and 3 units up. Place the location of his home on the grid and label the coordinate pair.

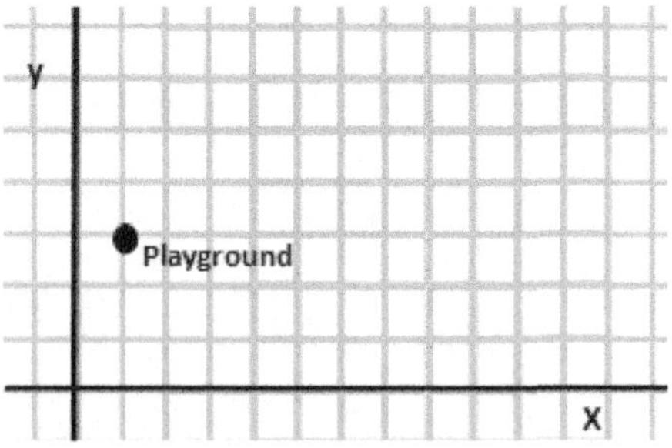

4. Select all triangles which are right triangles.

A. 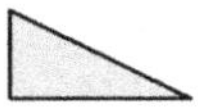

B. 

C. 

D. 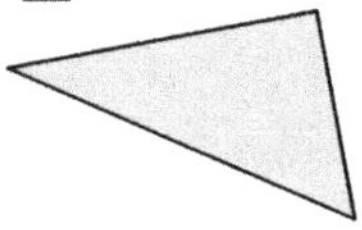

E. 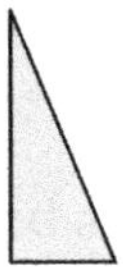

5. The volume of a cube is 27 $cm^3$. What are the lengths of the sides of the cube?

Need some hints? Look on page 186 for help

## Level 2 Geometry

1. What is the volume of the figure?

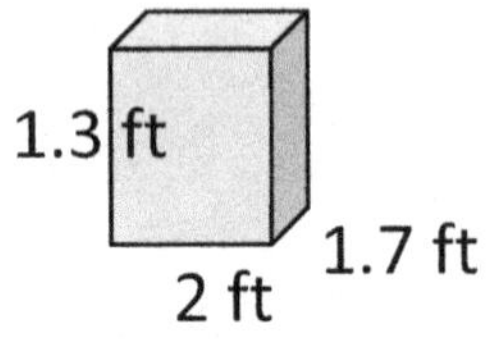


2. Trevor is buying mulch for a flower bed in his backyard. Each bag of mulch covers 3 square feet. How many bags of mulch will he need to buy?

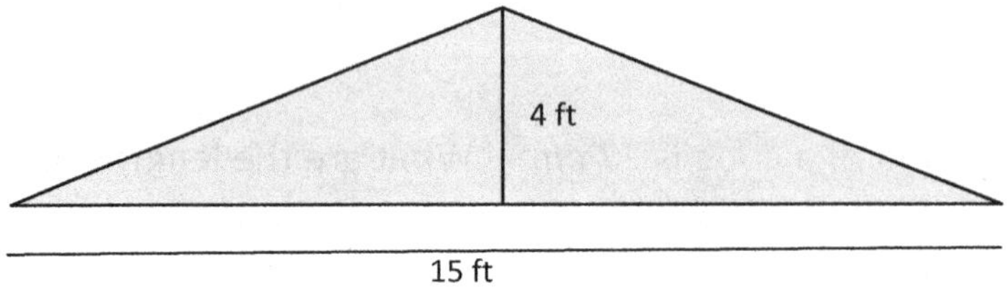


3. Find the area of the parallelogram.

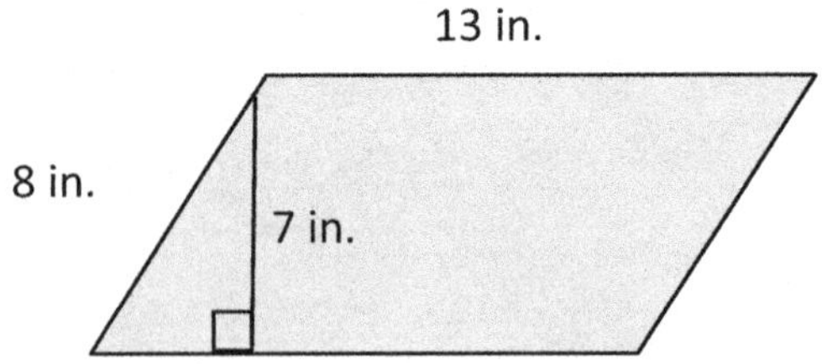

4. Two points are given on a coordinate plane. Add two additional points and connect to make a rectangle.

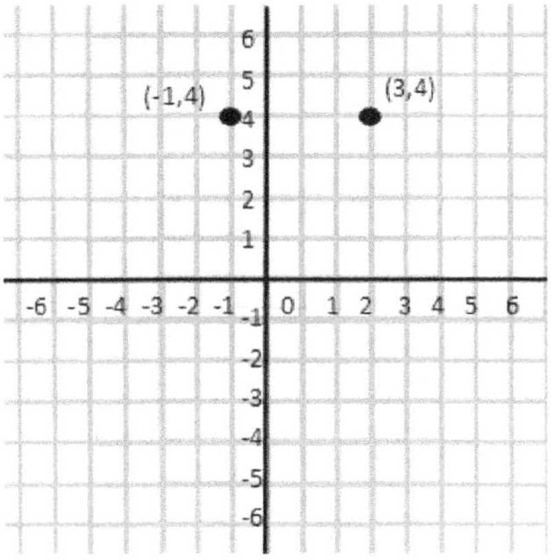


5. Sandy needed to wrap presents for her daughter's birthday. She had a roll of paper that will cover 25 $ft^2$. The dimensions of the boxes are below. Will Sandy have enough paper to cover the boxes? Explain your answer.

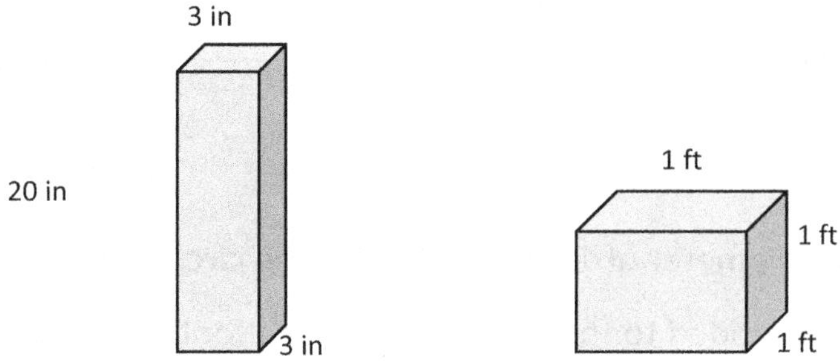


Need some hints? Look on page 186 for help

# Level 3 Geometry

1. A triangle has a side length of 5 inches and a side length of 2 inches. Which are possible lengths for the 3rd side of the triangle?

   A. 7 inches
   B. 3 inches
   C. 1 inch
   D. 4 inches
   E. 5 inches

2. The following 3 angles are supplementary angles. Find the missing angle.

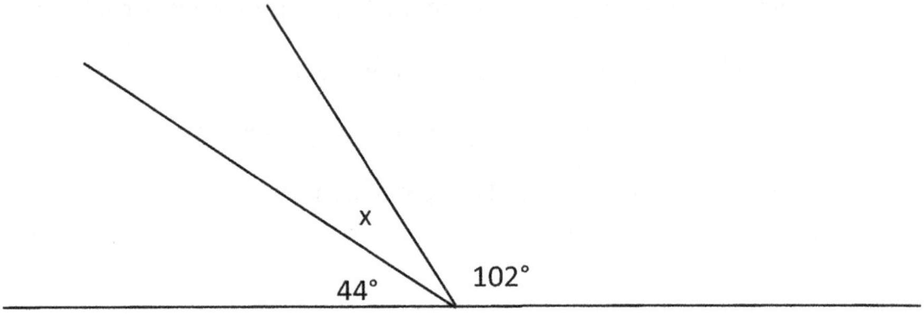


3. A circle has a diameter of 1.6 feet. What is the circumference of the circle, rounded to the nearest tenth of a foot? Use 3.14 for $\Pi$ .

4. Draw the cross section of the hexagonal prism if the cut is parallel to the base.

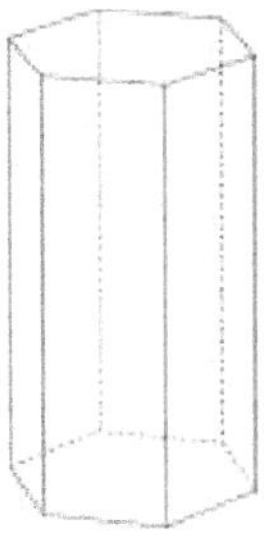

5. Two boxes have the same volume. One box has a base of 3 x 3 inches. The other box has a base of 6 x 6 inches. How many times as tall is the box with the smaller base?

Need some hints? Look on page 187 for help

## Level 4 Geometry

1. What is the volume of the cylinder?

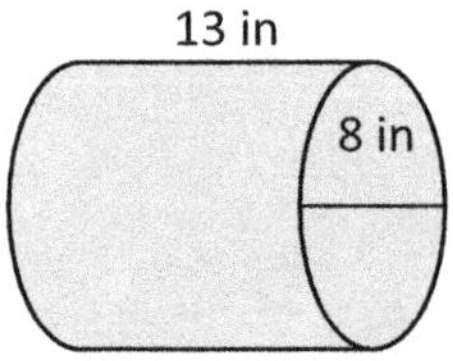


2. What is the measure of the missing angle if line segment CD is parallel to line segment EF?

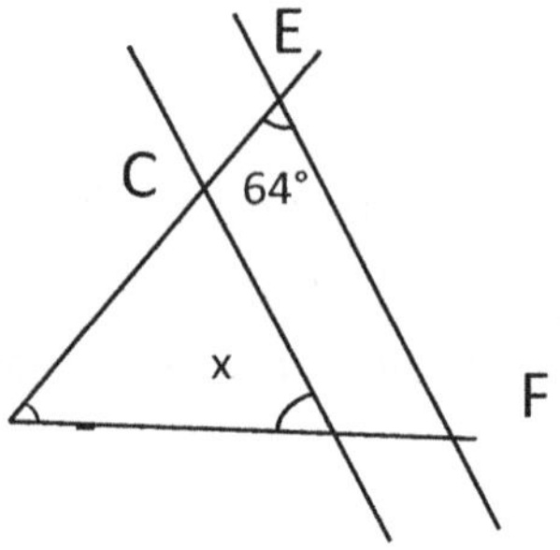


3. Angle GFI measures 35 degrees. What is the measure of angle GFH?

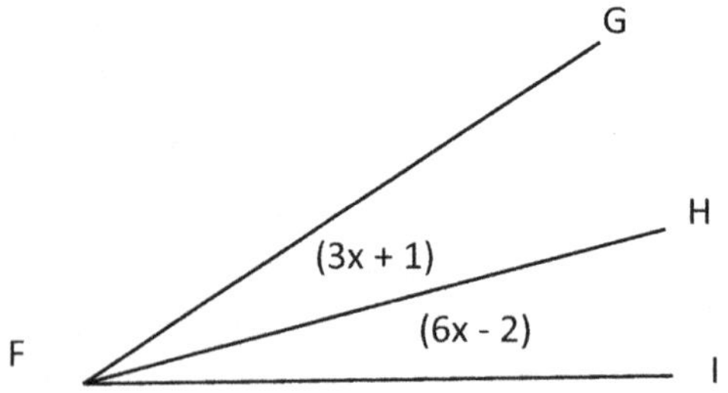

4. Is the movable figure congruent to $\triangle ABC$? If so, what transformations must occur?

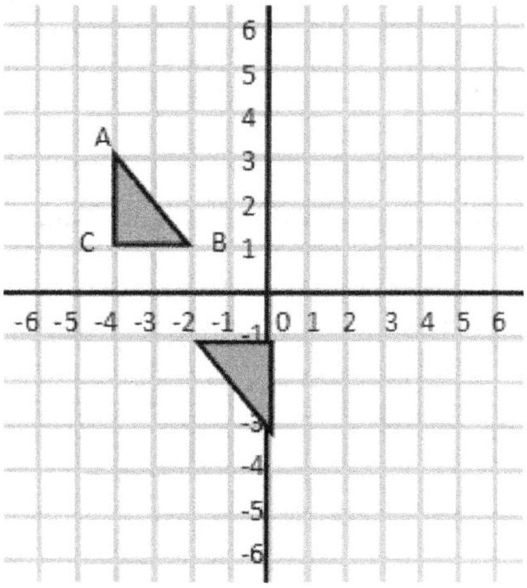


5. What is the distance between (3, -5) and (1, 4). Use the Pythagorean theorem.

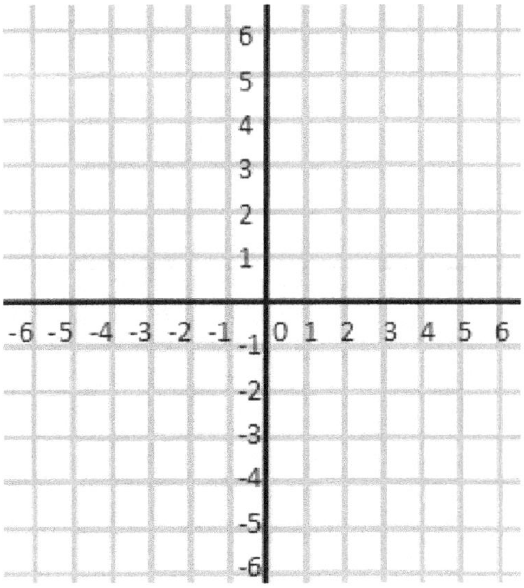


Need some hints? Look on page 187 for help

**SELF-ASSESSMENT ZONE**

**Level 1**

| Question | Answer | Correct? | Lucky Guess | Knew It | Simple Mistake | Need Practice |
|---|---|---|---|---|---|---|
| 1 | A, B, and D | □ | □ | □ | □ | □ |
| 2 | 42° | □ | □ | □ | □ | □ |
| 3 | (4, 6) | □ | □ | □ | □ | □ |
| 4 | A, D, and E | □ | □ | □ | □ | □ |
| 5 | 3 cm | □ | □ | □ | □ | □ |
| | **Total** | | | | | |

What do you think is going well?

What do you need to do to improve?

## SELF-ASSESSMENT ZONE
### Level 2

| Question | Answer | Correct? | Lucky Guess | Knew It | Simple Mistake | Need Practice |
|---|---|---|---|---|---|---|
| 1 | $4.42\ ft^2$ | ☐ | ☐ | ☐ | ☐ | ☐ |
| 2 | 10 bags | ☐ | ☐ | ☐ | ☐ | ☐ |
| 3 | $91\ in^2$ | ☐ | ☐ | ☐ | ☐ | ☐ |
| 4 | Any point for y as long as x stays at -1 for one ordered pair and for 3 other ordered pair | ☐ | ☐ | ☐ | ☐ | ☐ |
| 5 | No, she needs $27.5\ in^2$ | ☐ | ☐ | ☐ | ☐ | ☐ |
| | **Total** | | | | | |

What do you think is going well?

What do you need to do to improve?

## SELF-ASSESSMENT ZONE
### Level 3

| Question | Answer | Correct? | Lucky Guess | Knew It | Simple Mistake | Need Practice |
|---|---|---|---|---|---|---|
| 1 | D and E | ☐ | ☐ | ☐ | ☐ | ☐ |
| 2 | 34° | ☐ | ☐ | ☐ | ☐ | ☐ |
| 3 | 5 ft | ☐ | ☐ | ☐ | ☐ | ☐ |
| 4 | 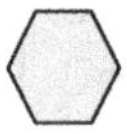 | ☐ | ☐ | ☐ | ☐ | ☐ |
| 5 | 4 times | ☐ | ☐ | ☐ | ☐ | ☐ |
| | **Total** | | | | | |

What do you think is going well?

What do you need to do to improve?

| SELF-ASSESSMENT ZONE<br>Level 4 | | | | | | |
|---|---|---|---|---|---|---|
| **Question** | **Answer** | **Correct?** | **Lucky Guess** | **Knew It** | **Simple Mistake** | **Need Practice** |
| **1** | $653.12\ in^3$ | □ | □ | □ | □ | □ |
| **2** | 59° | □ | □ | □ | □ | □ |
| **3** | GFH is 13° | □ | □ | □ | □ | □ |
| **4** | Yes, you would rotate and slide the figure | □ | □ | □ | □ | □ |
| **5** | $\sqrt{85}$ | □ | □ | □ | □ | □ |
| | **Total** | | | | | |

What do you think is going well?

What do you need to do to improve?

# SET A GOAL

*Adding a negative number always give you a lower sum. Negative behavior works the same way. Every bad habit we have takes away from our success. Write down one habit or behavior that might be holding you back, and then write down a goal for changing – just a little – in the next week. What will you do differently?*

# Chapter 7

## It really IS a small world

Some people have to learn how to fit in at a new school or in a new neighborhood. But what would happen if you had to move to a new *country?*

The authors of these stories all moved to new countries to chase their dreams. Read their stories to gain knowledge – and maybe a little inspiration.

**Weng Fatt Choy**

**Goodyear Employee Since 1995**

## Learning all new rules at work

*Weng Fatt Choy had to learn almost everything new when he moved to the United States. His kids adjusted quickly, but he needed help*

Transferring to Goodyear's global headquarters 12 years ago was a daunting task for both me and my family. I had to get used to a new country and a new work environment. My family and I had to adapt to new living arrangements in a different city.

*With great support from my teammates, I adapted to the new work culture.*

Although my kids had to enroll in a different school and make new friends, they were well adapted after one year. Adapting was not so easy for me.

I relied on my fellow team members to learn about my new position and responsibilities. My team taught me how I could best contribute to our work. With great support from my teammates, I adapted to the new work culture.

During this experience, I learned to trust the people I work with to help me. I also learned that fitting in may not always be easy for everyone, but it is always possible.

**FUN FACT:** Babies are all born able to learn any language in the world. After just 6 months, however, babies' ears only listen for the sounds of their own language. This is why adults from other countries who grew up speaking other languages have a hard time saying English words the "right" way; their brains haven't learned to make those sounds yet.

# REFLECTION ZONE

*If you had to move to a new country and learn a new language, how would you find your way? Write down one thing you would want to learn to say right away in that new language.*

**Christian Jurado**

**Goodyear Employee Starting in 2006**

## World traveler since 15

*Would you move to a new country alone? Christian did when he was 15, and he never regretted it*

As a child, I lived in a small town in the middle of the Austrian Alps, which is where I made all my close childhood friends. I went to the small school across the street and greeted every neighbor that crossed my path.

To this day, I can still recall the exact moment when my life changed forever. At the age of 15, on a Saturday afternoon in spring, my parents sat me down on the patio. I was given a choice: whether to stay in Austria and continue my life there, or depart as an exchange student to a country that only crossed my mind in geography class until that moment – the United States of America.

I recall my father eagerly telling my family in German, "This will be a great opportunity, and I believe that as long as you are capable of fitting in, you will be okay." I had desperately hung to those words, wanting to believe that it truly would be "okay".

*My culture is global, my traditions are diverse and my friends are from very different countries.*

And it turned out my father was right.

I knew I was going to end up sacrificing almost everything dear to me such as my friends, my school and my home. Now, more than 30 years later, I have had to fit in to new environments, cultures and life challenges.

I have moved to five different countries on three continents. I have learned four new languages, and have been exposed to different environments.

I know four things are important for mastering the challenge of fitting in. These qualities are: the ability to openly accept social differences without any prejudice; the ability to actively find new friends; the strength of your own judgment when in unfamiliar situations; and having faith in the support of your own family.

My culture is global, my traditions are diverse and my friends are from very different countries – all are results of my ability to fit in and accept the differences and challenges ahead.

**FUN FACT:** Kenneth Hale is recognized as the American who learned the most languages. Before he died, Hale had learned over 50 languages! He could pick up basic parts of a new language in 10 or 15 minutes.

# REFLECTION ZONE

*If you could learn any new language, what would it be? Why does that language interest you?*

**Dayang Nurttayati**

**Goodyear Employee Starting in 2013**

## From dreams to reality

*Dayang Nurttayati grew up with almost nothing, but she used her dreams as motivation for a better life*

When I was 10, I had a strong vision that one day I would be able to step into a university. Back then, my playground was alongside a "river" in an underdeveloped neighborhood. The "river" was the only place I could play outside. There was no colorful playground for me, and having a nice little doll was *impossible*.

One afternoon, a neighbor invited me over to watch a series on television. That little magic box truly inspired me. Seeing the world in the movie inspired me. I secretly promised myself that I would travel the world one day. *But how?*

*I kept imagining riding on an airplane to travel the world.*

I put in a little extra work on my homework the moment I reached home every day. I scored high on exams. Soon, I was at the top of my class.

My repeated silent prayers – "I will make it to the university one day" – kept playing in my head. I kept imagining riding on an airplane to travel the world.

That day eventually came when I was granted admission to The Flinders University in South Australia. I was timid then, but highly persistent. I made it to my big dream. It was truly a breakthrough!

My persistence carried me all the way to my dream, to where I wanted to be, and where I am today! Dream big. There is always a way to make your dreams come true!

**FUN FACT:** The busiest airport in the world is in Atlanta, Georgia, with over 101 million passengers every year.

# REFLECTION ZONE

*Dayang was inspired to see the world when she watched a TV show for the first time. Write down one thing that inspires you to chase your dreams.*

# Feeling inspired?

# LEVEL UP

**TRY THESE PRACTICE PROBLEMS**

Each level is a little more challenging

Check your work at the end

Look for hints if you need them

# CHALLENGE SET 7: NUMBER SENSE

## Level 1 Number Sense

1. Find all equivalent fractions to $\frac{2}{5}$ .

$\frac{1}{3}$ $\frac{12}{30}$ $\frac{10}{20}$ $\frac{6}{30}$ $\frac{8}{20}$ $\frac{20}{50}$

2. Solve $12.4 \div .4 =$

3. Write $5 \div 4$ as a fraction. Then, show this number on a number line.

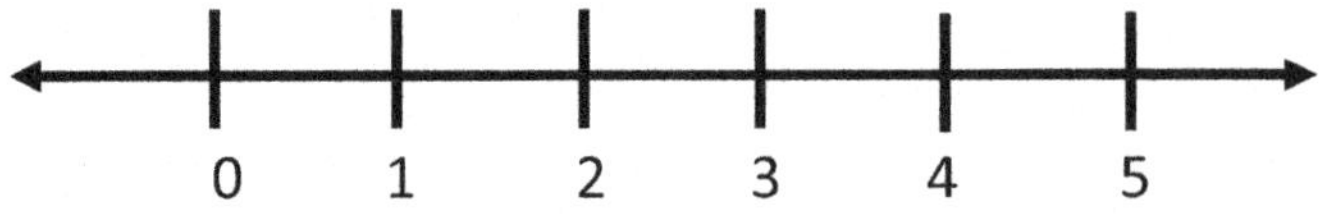

4. Dean wants to bring cookies to the party. The recipe makes 36 cookies. He needs 3 cups of sugar and 1 cup of butter. He only wants to make 12 cookies. How much butter and sugar will Dean need?

5. Michelle talked on the phone with her friends on Friday evening. She talked with Kayla for ¼ of an hour and to Sophie for ⅔ of an hour. What was the total amount of time Michelle was on the phone? How many minutes is this?

Need some hints? Look on page 188 for help

# Level 2 Number Sense

1. What is the greatest common factor of 8, 20, and 36?

2. Nate and Cole collected aluminum cans to recycle. Nate collected 1 ¾ lbs. and Cole collected 2 ⅛ lbs. of aluminum. They turned in the cans together. How many pounds of aluminum did they have in all?

3. Simplify the expression $\frac{1.6}{.5}$ .

4. What fraction(s) below are equivalent to $\frac{3}{8} \div \frac{1}{4}$?

$1\frac{1}{2}$ $\quad$ $\frac{3}{2}$ $\quad$ $\frac{12}{8}$ $\quad$ $1\frac{4}{8}$

A. $1\frac{1}{2}$ and $1\frac{4}{8}$

B. $\frac{3}{2}$

C. All of them

D. None of them

5. A serving of ice cream is ⅔ cup. Alexander got a belly ache because he ate 3 cups! How many servings did Alexander eat?

Need some hints? Look on page 188 for help

## Level 3 Number Sense

1. Simplify $\frac{8}{-11} + (\frac{-4}{5})$

2. If Andre wants to multiply -6 and a number to have a product that is a positive and less than 5, what could be a possible integer?

   A. 5
   B. $\frac{1}{5}$
   C. - 5
   D. $\frac{-1}{5}$
   E. $\frac{-1}{-5}$

3. Simplify $(-3)^2 + \frac{1}{5} - (-3.2)$

4. The submarine was returning to the surface dock and the top of the vessel went from 86 ft below sea level to 12 feet above sea level. What is the total distance, in feet the submarine traveled to return to the surface dock?

5. What is the value of the expression $\frac{\frac{5}{6}+\frac{1}{2}}{\frac{3}{8}}$?

Need some hints? Look on page 189 for help

## Level 4 Number Sense

1. Which number(s) are perfect squares?

   A. 64
   B. 144
   C. 110
   D. 6
   E. 9

2. A star was measured at a distance of 3,110,000 miles away from earth. Write this number in scientific notation.

3. Which of the following describes the square root of 47?

   A. Between 5 and 6
   B. Between 46 and 48
   C. Between 6 and 7
   D. Between 22 and 25

4. Simplify the following expression $\frac{1}{4}(\sqrt{25 \cdot 4} + 14)$ .

5. Miguel was solving the problem below.

$$\frac{1}{4}(\sqrt{16}) - \frac{2 \div 1/2}{2^{-2}}$$

$$4 - \frac{2 \div 1/2}{\frac{1}{4}}$$

$$4 - \left(\frac{4}{\frac{1}{4}}\right)$$

$$4 - 16$$

$$-12$$

What error did Miguel make? What is the correct answer?

Need some hints? Look on page 189 for help

| SELF-ASSESSMENT ZONE<br>Level 1 | | | | | | |
|---|---|---|---|---|---|---|
| **Question** | **Answer** | **Correct?** | **Lucky Guess** | **Knew It** | **Simple Mistake** | **Need Practice** |
| 1 | $^{12}/_{30}$ $^{8}/_{20}$ $^{20}/_{50}$ | ☐ | ☐ | ☐ | ☐ | ☐ |
| 2 | 31 | ☐ | ☐ | ☐ | ☐ | ☐ |
| 3 | 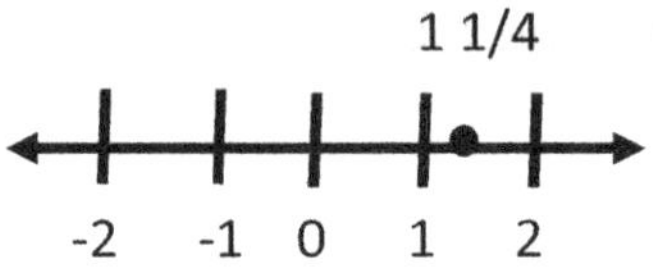 | ☐ | ☐ | ☐ | ☐ | ☐ |
| 4 | 1 c. sugar and $^{1}/_{3}$ c. butter | ☐ | ☐ | ☐ | ☐ | ☐ |
| 5 | $^{11}/_{12}$, 55 minutes | ☐ | ☐ | ☐ | ☐ | ☐ |
| | **Total** | | | | | |

What do you think is going well?

What do you need to do to improve?

| SELF-ASSESSMENT ZONE Level 2 | | | | | | |
|---|---|---|---|---|---|---|
| Question | Answer | Correct? | Lucky Guess | Knew It | Simple Mistake | Need Practice |
| 1 | 4 | ☐ | ☐ | ☐ | ☐ | ☐ |
| 2 | $3\ ^{7}/_{8}$ lbs. Aluminum | ☐ | ☐ | ☐ | ☐ | ☐ |
| 3 | 3.2 | ☐ | ☐ | ☐ | ☐ | ☐ |
| 4 | C | ☐ | ☐ | ☐ | ☐ | ☐ |
| 5 | $4^{1}/_{2}$ | ☐ | ☐ | ☐ | ☐ | ☐ |
| | Total | | | | | |

What do you think is going well?

What do you need to do to improve?

| SELF-ASSESSMENT ZONE<br>Level 3 | | | | | | |
|---|---|---|---|---|---|---|
| Question | Answer | Correct? | Lucky Guess | Knew It | Simple Mistake | Need Practice |
| 1 | $1\,^{29}/_{55}$ | □ | □ | □ | □ | □ |
| 2 | D | □ | □ | □ | □ | □ |
| 3 | 12.4 or $12\frac{2}{5}$ | □ | □ | □ | □ | □ |
| 4 | 98 ft | □ | □ | □ | □ | □ |
| 5 | $3\,^{5}/_{9}$ | □ | □ | □ | □ | □ |
| | Total | | | | | |

What do you think is going well?

What do you need to do to improve?

| SELF-ASSESSMENT ZONE<br>Level 4 | | | | | | |
|---|---|---|---|---|---|---|
| **Question** | **Answer** | **Correct?** | **Lucky Guess** | **Knew It** | **Simple Mistake** | **Need Practice** |
| **1** | A, B, and E | ☐ | ☐ | ☐ | ☐ | ☐ |
| **2** | $3.11 * 10^6$ | ☐ | ☐ | ☐ | ☐ | ☐ |
| **3** | Between 6 and 7 | ☐ | ☐ | ☐ | ☐ | ☐ |
| **4** | 6 | ☐ | ☐ | ☐ | ☐ | ☐ |
| **5** | $1/4\sqrt{16} = 1$, not 4, the answer is -15 | ☐ | ☐ | ☐ | ☐ | ☐ |
| | **Total** | | | | | |
| What do you think is going well? | | | | | | |
| What do you need to do to improve? | | | | | | |

# SET A GOAL

*Did you know that your brain **changes** when you try something new? If you are right-handed and you start writing with your left hand, your brain will change to build new connections for left-handed writing. Set a goal to try something new and challenging in the next week to grow a stronger brain.*

# Chapter 8

## HELP is a gift we can always ask for

No matter who you are, or how successful you become, there will always be a moment when you need to ask for help. For some people, asking for help can be very hard. They may be shy, or they might think that asking for help makes them look stupid.

It turns out that asking for help once in a while is the smartest thing you can do. We can't be right all the time, and we can't always figure things out on our own.

The authors of these stories learned to ask for help when they were having a hard time in school. And because they were brave enough to ask for help, they were a lot more successful.

**Karen Burke**

**Goodyear Employee Starting in 1984**

## Failure can be the first step to success

*Karen didn't let a big setback scare her away from learning math. Now math is a part of her career.*

Think of any classroom, but particularly a math class. Imagine all the students settle down as class is about to begin, and then the teacher says... "Pop quiz!"

I began to sweat as my teacher passed out the quiz. I looked at problem No. 1... *and didn't know how to solve it*. Problem No. 2? No clue...

*I thought to myself, "I can figure this out. I know I can do this, especially if I ask for help from my math teacher."*

I started working on the problems, trying to figure them out. I dragged my feet as I turned in the quiz.

Fast forward to the next day, when my teacher returned my quiz to me with a big, fat "F" on it. As you can imagine, I was very upset. But once I calmed down, I thought to myself, "I can figure this out. I know I can do this, especially if I ask for help from my math teacher."

Knowing that help was available inspired me to come back from this "F," ask for help from my teacher, and work really hard in math. I eventually came back from that defeat to excel at math and even participate in honors math courses. Eventually I used my math and science excellence to pursue a career in engineering.

Don't let one failure define you. Learn from it. Ask for help, and work hard to fix the issue.

**FUN FACT:** Popcorn pops at around 356° Fahrenheit (180° C). Popcorn might be one of the oldest snacks; some pieces of popcorn have been found that are over 4,500 years old!

# REFLECTION ZONE

*Write down one area in math where you know you need a little help. Then write down a plan you have to get the help you need.*

**Megan Clutter**

**Goodyear Employee Starting in 2014**

## Smart is something you become

*Megan worried that asking for help would make her look stupid. Getting over that fear helped her succeed*

When I was little, I always loved school. The thought of learning something new was always exciting to me, and learning came easy. But as I got older, school started to get harder. I always feared people wouldn't think I was smart anymore if I asked a silly question in class.

So, I kept quiet because I was scared of what others would think. I tried studying as hard as I could with my books, but I still got lost.

Then one day, after a really confusing class, I walked up to my teacher and asked for help with understanding the topic for the day. To my surprise, she was happy I asked! She brought me to her office, and we worked through the problem in a different way. Suddenly, what I thought was extremely hard became extremely clear.

The next class with that same teacher had reached a point where I was confused again. I started to raise my hand, but quickly brought it down because I felt nervous. We weren't in the comfort of her office this time, and I was now in a class of 30 people. My teacher had noticed this, and gave me a small nod that meant, "It's okay to ask."

*I kept quiet because I was scared of what others would think.*

I put my hand back up, asked my question and braced myself for being judged. Instead, the girl who sat next to me whispered, "I had the same question, but didn't want to ask!" I felt relieved. I wasn't alone. More importantly, I stopped being afraid to ask questions.

To this day, I can happily say I have never had a manager, teacher or coworker who has turned me away for asking for help or wanting to learn something new. Sometimes you just need to be willing to ask!

**FUN FACT:** 100% of students will be more successful if they ask for help from a teacher or others who can explain what they don't understand!

# REFLECTION ZONE

*Write down how you will ask for help the next time you do not understand something in math class.*

*Feeling inspired?*

# LEVEL UP

**TRY THESE PRACTICE PROBLEMS**

Each level is a little more challenging

Check your work at the end

Look for hints if you need them

# CHALLENGE SET 8: DATA AND STATISTICS

## Level 1 Data and Statistics

1. What type of graph would be the best choice to display the results of a poll for what month people prefer to vacation in Florida?

   A. circle graph
   B. stem and leaf
   C. histogram
   D. line plot

2. The Badger baseball team kept track of all doubles hit in the season each game. How many total doubles were hit by the Badgers in all the games played?

3. The table shows the temperatures for California in the month of July.

| Temperature | 70 -74 | 75 - 79 | 80 -84 | 85 -89 |
|---|---|---|---|---|
| Days | 5 | 14 | 8 | 4 |

How many days was the temperature below 85 degrees?

4. Check the appropriate probability box for each event to occur:

| | No chance | Unlikely | Likely | Certain |
|---|---|---|---|---|
| A. The day after Saturday will be Sunday | | | | |
| B. You will need a winter coat in Florida | | | | |
| C. The sun will be up at 1:00 AM in Ohio | | | | |
| D. Dinner will be served between 5-7 PM | | | | |

5. A study was conducted on what kind of pie people prefer. The study showed that 50% of people prefer apple, 30% prefer cherry, and 20% prefer blueberry. Fill in the pie chart with the given information. Draw the dividers and label each section appropriately.

Need some hints? Look on page 190 for help

## Level 2 Data and Statistics

1. How many more squares and circles are needed to make the ratio of squares to total be 4:13?

2. The weights of 2-week-old puppies born to a golden retriever are given in the chart below.

| | | | | |
|---|---|---|---|---|
| 4.2 | 4.6 | 3.9 | 3.9 | 4.0 |
| 4.2 | 3.9 | 3.8 | 3.1 | |

What is the mean (to the nearest tenth), median, and mode for the weights of the puppies?

3. John rolled a 6-sided die 20 times. About how many times would you expect John to roll an even number? Explain why you think this.

4. A homeroom teacher wants to learn about her class. Which question below would be a good research question to ask?

    A. What is your name?
    B. What is your favorite subject?
    C. Who lives at 123 West Maple Street?
    D. Who turned off the lights?

5. Create a dot plot (line plot) for the number of points scored by players on the girls basketball team during Thursday's game.

    2, 1, 2, 7, 10, 7, 4, 5, 2, 8, 2, 4

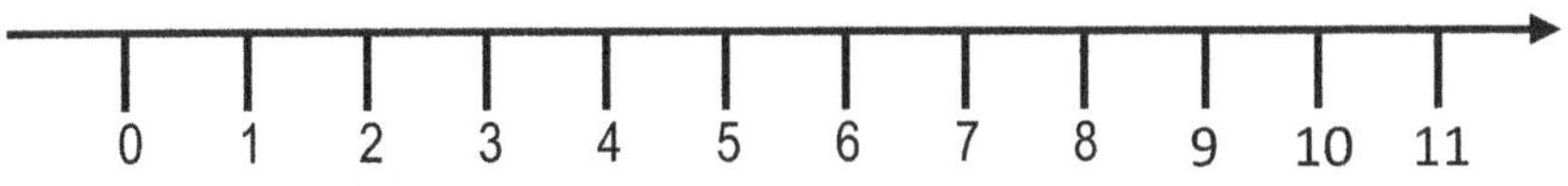


What is the range for the points scored? ________

How many girls scored less than 6 points?________

What is the total number of points scored by the team?________

Need some hints? Look on page 190 for help

# Level 3 Data and Statistics

1. Which event is least likely to occur?

   A. Spinning a spinner divided into 3 sections labeled 1-3 and landing on a number greater than 1
   B. Flipping a coin with sides of heads or tails, and landing on heads
   C. Rolling a number cube with faces labeled 1-6, and the cube landing on an even number
   D. Pulling the eight of clubs card randomly from a standard deck of 52 cards

2. A student is randomly selected from a class of 28 students. What is the probability a boy will be selected if there are 16 girls in the class?

3. Estimate the number of times you need to roll a dice to get the number five as the outcome 6 times. How many times would you have to roll the dice to get a five or six as the outcome 6 times?

4. The carnival has a game set up to roll a dice and spin a spinner like the one shown here.

What is the probability of rolling a three on the dice and spinning a three on the spinner in the same turn?

5. The ratio of cats to dogs at the kennel is 4:7. Each week, there are about 28 cats at the kennel. About how many dogs are at the kennel each week?

Need some hints? Look on page 191 for help

## Level 4 Data and Statistics

1. Which situation best describes the information displayed on the scatterplot?

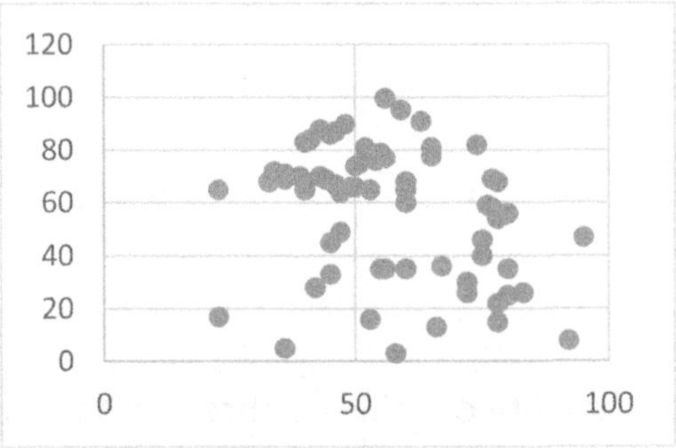


A. Positive correlation
B. Negative correlation
C. No correlation
D. Positive nonlinear correlation

2. Select all data sets which have a range larger than 5.

A. [2,3,6,2,5,4,6,3,4]
B. [6,7,6,1,4,6,7,4,5,6]
C. [3,4,7,9,3,4,7,9,3,4,7,9]
D. [1,1,2,1,1,1,0,1,2,0]
E. [8,10,16,11,7,6,8,12,10,7]

3. Draw the line of best fit on the scatter plot below

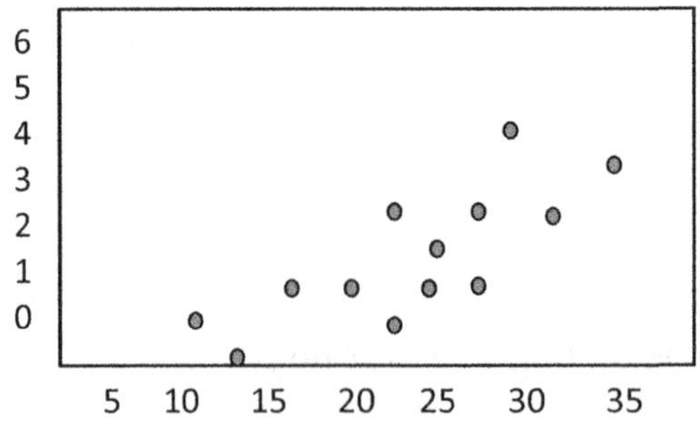

4. Anikate has 8 coins in his pocket, 2 quarters, 3 dimes, and 3 nickels. He pulls a coin at random, looks at it, and puts it back. He selects another coin at random from the same pocket. What is the probability Max selected a dime or a nickel both times?

5. At the ice cream social, kids could have toppings put on of sprinkles, chocolate syrup, or both.

- A total of 56 children chose both
- A total of 101 used chocolate syrup
- A total of 78 used sprinkles
- There was a total of 133 children in attendance

Complete the Venn Diagram to display how many chose just sprinkles and how many chose just chocolate syrup.

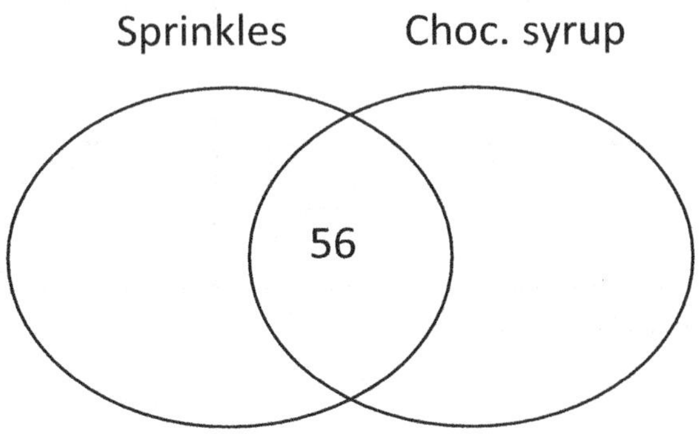


Need some hints? Look on page 191 for help

## SELF-ASSESSMENT ZONE
### Level 1

| Question | Answer | Correct? | Lucky Guess | Knew It | Simple Mistake | Need Practice |
|---|---|---|---|---|---|---|
| 1 | D. A line plot would show how many people pick each month, and then you identify the month with the most. | ☐ | ☐ | ☐ | ☐ | ☐ |
| 2 | 53 doubles | ☐ | ☐ | ☐ | ☐ | ☐ |
| 3 | 27 | ☐ | ☐ | ☐ | ☐ | ☐ |
| 4 | A. Certain<br>B. Unlikely<br>C. No chance<br>D. Likely | ☐ | ☐ | ☐ | ☐ | ☐ |
| 5 | 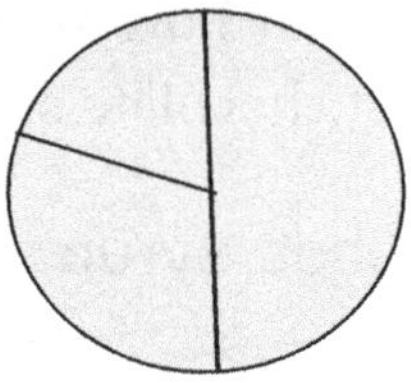 | ☐ | ☐ | ☐ | ☐ | ☐ |
| | Total | | | | | |
| What do you think is going well? | | | | | | |
| What do you need to do to improve? | | | | | | |

# SELF-ASSESSMENT ZONE
## Level 2

| Question | Answer | Correct? | Lucky Guess | Knew It | Simple Mistake | Need Practice |
|---|---|---|---|---|---|---|
| 1 | Squares = 0<br>Circles = 3 | □ | □ | □ | □ | □ |
| 2 | Mean 4, Median 3.9 and Mode 3.9 | □ | □ | □ | □ | □ |
| 3 | About 10 times. Half the numbers on a 6-sided die are even. | □ | □ | □ | □ | □ |
| 4 | B | □ | □ | □ | □ | □ |
| 5 | range: 9 points<br>8 girls scored less than 6 points<br>the team scored 54 points total | □ | □ | □ | □ | □ |
| | **Total** | | | | | |

What do you think is going well?

What do you need to do to improve?

| SELF-ASSESSMENT ZONE<br>Level 3 | | | | | | |
|---|---|---|---|---|---|---|
| Question | Answer | Correct? | Lucky Guess | Knew It | Simple Mistake | Need Practice |
| 1 | D | □ | □ | □ | □ | □ |
| 2 | 3/7 chance a boy is chosen | □ | □ | □ | □ | □ |
| 3 | 36 times for a 5<br>18 times for a 5 or 6 | □ | □ | □ | □ | □ |
| 4 | $^1/_{18}$ | □ | □ | □ | □ | □ |
| 5 | 49 dogs. $\frac{4}{7} = \frac{28}{49}$ | □ | □ | □ | □ | □ |
| | Total | | | | | |

What do you think is going well?

What do you need to do to improve?

## SELF-ASSESSMENT ZONE
### Level 4

| Question | Answer | Correct? | Lucky Guess | Knew It | Simple Mistake | Need Practice |
|---|---|---|---|---|---|---|
| 1 | C | ☐ | ☐ | ☐ | ☐ | ☐ |
| 2 | B, C and E | ☐ | ☐ | ☐ | ☐ | ☐ |
| 3 | 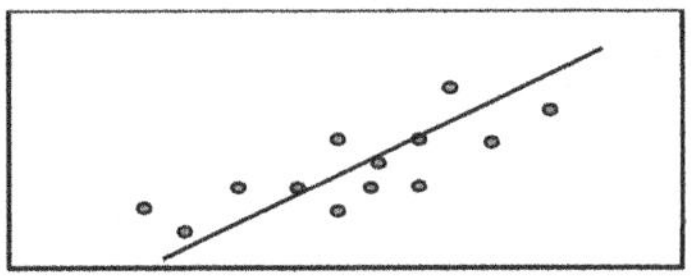 | ☐ | ☐ | ☐ | ☐ | ☐ |
| 4 | $^{9}/_{16}$ | ☐ | ☐ | ☐ | ☐ | ☐ |
| 5 | 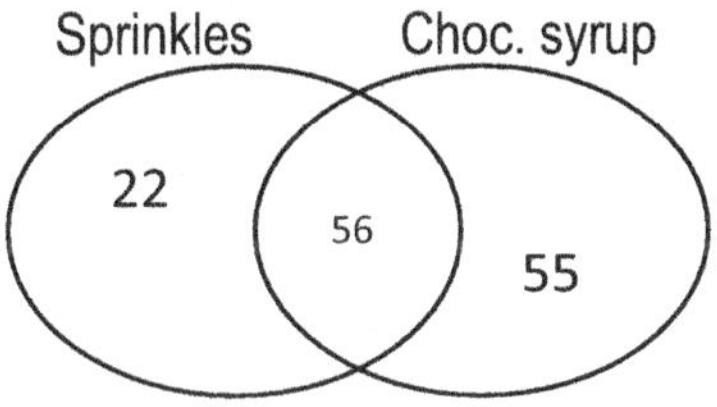 | ☐ | ☐ | ☐ | ☐ | ☐ |
| | **Total** | | | | | |

What do you think is going well?

What do you need to do to improve?

# PAT YOURSELF ON THE BACK

*You did it! Maybe these challenge problems were hard for you. Or maybe they were easier than you expected. Either way, you stuck with it to the end – and that will make you stronger in the future.*

*Write a letter to the NEXT student who gets a copy of this book. What advice would you share? How would you encourage that student to stick with it and improve? We will help you start.*

Dear student,

# Hints

## Chapter 1 Algebra

Level 1

1. Don't forget PEMDAS (do what's in the parentheses first)

2. Look for keywords. "More than" means addition. "Times" means multiplication.

3. What happens if you substitute a "0" in for b in the equation?

4. Look for a constant change between the in and out table. If the input is 1 and the output is 4, then it could be increasing by 3 or multiplying by 4. Then check the possibilities with another function set. If it works, you found the pattern.

5. See if you can find a way to figure out how much the ribbon is for each yard.

Level 2

1. What does the exponent mean? The 6th power does not mean times 6, but represents what in relation to the base number?

2. Look for the key math words in the problem. Increase is adding, each is dividing, what buzz words can you find to tell you the operation?

3. This requires the use of the distributive property. The word distribute means to 'hand out' or 'pass out' the number outside the parentheses.

4. We know the number of boys the same as the number of girls plus 160 more. Use x to represent the number of girls. What would you use to represent the boys?

5. Look for the information that will not change. This will be the constant. Remember that the MOST he can spend is \$80, so that means it can be \$80, but not \$86.

Level 3

1. Go through the words and put the math symbol above its written word (ex. 7 is seven, = is equal)

2. Use different grouping symbols to find all the 'like terms' and their integer signs. Use squares, circles, or triangles to visually see which terms to combine.

3. Think about the key word per. What is the problem telling us changes? That will become the variable.

4. To find out his salary for a month we would divide by 12, since there are 12 months in a year. What should we use to find out weekly? Biweekly?

5. Use a variable to represent the first number in the sequence. The problem states the numbers are consecutive which means they are in order from one to the next such as 5, 6, 7, 8. Look at how the numbers relate to the first number in this sequence. Now apply that same idea to the problem.

Level 4

1. Remember the rules associated with the operations of powers.

2. Because it is multiplying, we will add the exponents.

3. This is when we need to use what we know in the equation. Try substituting in a few numbers for x and look at what happens. Also, look at what happens to the value of x every time.

4. Find out what the graph is showing for the relationship between days and lawns first.

5. Get like terms gathered on one side and solve for x. This will show you what the outcomes are based on your findings.

## Chapter 2 Geometry

Level 1

1. What do we know about parallelogram? We also know that all 4 sides of a parallelogram have to equal 360 degrees.

2. Make a list of what you know is true of the properties of a square, then see what a parallelogram shares.

3. Count up how many cubes make up each dimension of the box. What do you do once you find the dimensions? Area is length x width of a 2-dimensional figure we want to fill, but the box is 3-dimensional figure we want to fill. So, what can we do?

4. Remember he wants to get home from the store, so start with your finger or pencil at that location and follow directions to get home.

5. Volume is length x width x height, we know the total volume and we know the height is 3. If we can figure out what number times 3 gives us 48, then we can figure out the other two sides.

Level 2

1. We can use what we know about distance here. Distance is the amount of space between to points.

2. Start with writing the dimensions for each individual rectangle. Then you can find the areas for each individual rectangle that makes up the prism.

3. Estimate an answer first, that way you will know you have your decimal right. Remember when multiplying by a decimal you put the decimal back in as many places as there are numbers on the right of all decimals.

4. Put the integers in order on the paper first. See how many should be on the left of the zero and how many on the right of the zero.

5. A cube is a 3 dimensional figure with square faces. This also means all sides are equal.

Level 3

1. Remember that in order for a triangle to form, two sides together must be longer than the third side so when all sides touch, they form the peak of the triangle.

2. A cross section is like a slice. It is the 2-dimensional shape that results from slicing a 3-dimensional figure. Think of when you slice an orange. You cut it and pull the pieces apart; the inside looks like a circle.

3. Remember to use Pi to solve this equation.

4. Two important pieces of information we must know is that complementary, in this case, means equal to 90 degrees and that all vertical angles (angles opposite of each other) are equal.

5. The circumference is given so that means we know the diameter. For area we need the radius. Think about the relationship between the diameter and radius, then use the formula for the area of a circle.

Level 4

1. You will have to find the volume of both cylinders first ($\Pi r^2 h$). Find the math buzz word in the problem to help you figure out which math operation is needed to solve the problem.

2. The two angles together have to equal 180 degrees. So, you can set up the equation as y + 2x + 6 = 180.

3. This looks like a right triangle based on the diagram, so you will need the Pythagorean theorem.

4. Take this one movement at a time, draw in lightly what happens first and follow the measurement until the last step.

5. If we dilate an object we are "zooming in" on it.

## Chapter 3 Number Sense

Level 1

1. Can you simplify any of the fractions? Equivalent fractions are the same ratio between the numerator and denominator.

2. What does "quarter" mean? How many quarters are in a whole?

3. When a fraction is larger in the numerator, it is improper. Can you turn it into a mixed number by seeing how many of the denominators you can take out of the numerator?

4. How much further is the same as saying "what is left". Remember the keywords to help you decide what operation to use.

5. Figure out some equivalent fractions to $\frac{1}{2}$ .

Level 2

1. Remember making factor trees? See what is the largest number they all share.

2. First, figure out what fraction is the problem representing....is there another way we could write that fraction which means the same amount?

3. Remember when dividing by a decimal you are actually making more of something so your answer should be larger than the dividend.

4. The problem tells us how much they each ate of the pizza, but we need to know how much is left. Figure that out for each person first. Then think of the key word together in the problem...what operation is that?

5. If one serving is $\frac{2}{3}$ cups and we need 2 cups, how many $\frac{2}{3}$ can go into 2? What operation is this asking? Or we could look at 2 cups as an improper fraction. How the improper fraction compares to the original serving size?

## Level 3

1. You cannot add pieces that are not the same size. Make sure the fractions have common denominator and use the integer rules.

2. Try substituting in a number and see what criteria fits the answer.

3. Remember PEMDAS and the rules for adding/subtracting integers (ie. KCC-keep change change)

4. Changing the decimals to fractions or fractions to decimals will set the problems up for solving them.

5. We know the average for the week is 3, and we know there are 7 days in a week. So what should the total of the temperatures added together be? Now remember you need to fill in 2 numbers to make the total be the number you want for the temperatures added.

## Level 4

1. Think of what the terms mean and how the fraction relates. Another name for a fraction is a ratio.

2. Scientific notation is a way to express really large or really small numbers using a base 10 and an exponent. A positive exponent makes a large number and a negative exponent makes a really small number. The exponent tells you the direction to move the decimal.

3. PEMDAS (remember a square root is the inverse of an exponent so you complete a square root in the same place as the E in PEMDAS)

4. Dividing a number by a whole number will create a smaller number, what happens with an integer? A fraction?

5. Fractions, decimals, and percents are all interchangeable. Set the information in a fraction and then you can find the decimal and percent equivalent.

## Chapter 4 Data and Statistics

Level 1

1. The graph needs to display a change over time.

2. Each section is for a group of wrestlers for specific weights. We want the total. What operation does total mean?

3. When making a line plot, each dot you make represents one number in the data set. For instance, if there are three 7's, you need 3 dots above the number seven.

4. This is the same as a part of a whole where the "whole" is the total amount of numbers in the set.

5. Since this is a double bar graph, we need to compare two things. Think of different ideas where there may be two opinions or two different sets of data to mark and compare over a duration of time.

Level 2

1. Since we know we are looking to add to the jar, look first at the ratio that actually does exist in the jar, then you can decide what you need to add.

2. Statistical answers are answers that do not have one specific answer to the question. It could be a preference or choice.

3. Mean = average; median = middle; mode = most

4. Histogram are much like a bar graph, however, the data is grouped together in ranges to display the results.

5. Stem and leaf plots are easy to analyze if you can remember that the first column represents the "tens" place value, making the stem, and the leaf side is the "ones" place value.

Level 3

1. A true random sample means you would poll or survey an unbiased set of people. Meaning, the opinion does not belong to a specific group of individuals.

2. We first have to recognize the original probability of rolling a two. We can then use this number to figure out the question.

3. Since we know there are 100 total gumballs, this means each kind has a certain number in the machine out of 100. Set up as fractions to solve the problem.

4. Most likely means the majority of the time or should happen often.

5. In this problem, we know the amounts for the sample selected out of 80, since we know the total students, we should be able to set these up as proportions or equivalent fractions to find the answer.

Level 4

1. The direction the dots are heading can give us a clue to what the answer will be. Are they rising, falling, or not changing?

2. To answer this problem, we need to remember what an outlier is. It is an 'outside' number, or a number that does not follow the trend of the rest of the dots.

3. The range is the difference between the highest and lowest number.

4. We can first find out the probability of it happening one time, then we know we are using the same probability for the second. When we want to know the probability of both we need to combine the two with multiplication.

5. There are buzz words we need to listen for as we read. 120 roses are 60% OF the total. Of means multiply, so we can set up an equation and solve for 'x' to know how many total flowers there were. We then can answer the original question asked in the problem.

## Chapter 5 Algebra

Level 1

1. Use order of operations (PEMDAS).

2. The math buzz word here is quotient. We also need to look at that the second number has its own operation to handle BEFORE we can use it to find the quotient.

3. There are two separate sequences to follow. Treat each one individually to fill in the chart.

4. Remember PEMDAS here to help you see what operation you need to perform with the missing number. This should help you shorten the range of numbers to guess and check.

5. Look for buzz words that match the expression and the numbers used in the word problem match the expression completely.

Level 2

1. Two things we should focus on here. Remember order of operations (PEMDAS) and that a number squared is the same as multiplying that number times itself.

2. With inequalities, we are looking for the buzz words to show us if we should use greater than, less than, greater or equal, or less than or equal.

3. Factoring out a number here tells us we need to “remove” that number from each term in the expression.

4. Set up the problem with b = books, meaning any book and make an expression representing fiction and nonfiction in terms of *b*. This way you can combine terms and solve as a 2-step equation for *b*.

5. If each hour he earns $8.75, how do we find out how much for 2 hours? 3 hours? What operation would this require?

Level 3

1. Since it asks us for the product, we will need to use the distributive property first.

2. If we can see all the terms separately in the expression, we can see which of the options are combinations that include the terms. Use distributive property first, then analyze the options.

3. If we can identify the constant and the variable here, we can set up a 2-step equation to solve for "h."

4. Remember that the original price is the same as paying 100% of the cost. So, we can take the discount percent off of the total percent of the cost first before solving as one method.

5. Use a variable to represent the first number in the sequence. The problem states the numbers are consecutive which means they are in order from one to the next such as 5, 6, 7, 8. Look at how the numbers relate to the first number in this sequence. Now apply that same idea to the problem.

Level 4

1. Remember the rules associated with the operations of powers. Also remember that scientific notation is written with a number between 1 and 10. We may need to increase or decrease a power to make this happen.

2. Because it is multiplying, we will add the exponents.

3. If we use distributive property first, we will be set up to combine like terms.

4. We need to recall that for relationships to be a function, there must be 1 output for every input.

5. Before we can set up an equation, we need to figure out how much a ticket is. We can use two different data sets from the table to compare how the price changed and how many more tickets were sold. Then we can figure out the price per ticket and continue solving the problem.

## Chapter 6 Geometry

### Level 1

1. Make a list of what you know is true of the properties of a trapezoid, then see what a parallelogram shares.

2. The trapezoid has a line of symmetry down the middle, that means each side has what kind of relationship to each other?

3. Remember he wants to get home from the playground, so start with your finger or pencil at that location and follow directions to get home.

4. This question is looking at the comparison of angles on a triangle for classification. There are acute, obtuse, and right triangles. Recall the definition for each and see if we can rule out any choices.

5. We know that to find the volume we need to multiply length x width x height. Since we know a cube is the 3-dimensional version of a 2-dimensional shape we can look at the faces of the cube to determine what the shape is. This will help decide the relationship between the edges of the cube.

### Level 2

1. Volume formula is l x w x h .

2. The garden is in the shape of a triangle, so we must first figure out the total area we need to cover.

3. A parallelogram is really like a rectangle that is stretched to one side or the other. The distance between the top and bottom line of the parallelogram does not change. So, if we can recognize the height of the parallelogram (going straight up) we can use the formula for the area of a rectangle to achieve the same answer.

4. The points are in line horizontally, so as long we keep the next two points parallel at a 90-degree angle, we should get a rectangle.

5. This requires a few steps. We need to know the total volume of both boxes and we need to convert between inches and feet to keep our units of measurement the same. Recall there are 12 inches in a foot.

Level 3

1. Remember that in order for a triangle to form, two sides together must be longer than the third side so when all sides touch, they form the peak of the triangle.

2. Supplementary angles must add up to 180 degrees.

3. The distance around a circle is always $\Pi$ times the diameter.

4. A cross section is like a slice. It is the 2-dimensional shape that results from slicing a 3-dimensional figure. Think of when you slice an orange. You cut it and pull the pieces apart; the inside looks like a circle.

5. You can set this up as an equation where the two figures are equal to each other. For instance, we know that L x W x H is used for volume, we know the lengths and widths of both. So, we can set them up equal using a different variable for the height. Then look at the relationship when you substitute a number in for one of the variables.

Level 4

1. The volume of a cylinder ($\Pi r^2 h$).

2. The shape actually forms two triangles that are congruent. What do we know about congruent triangles?

3. Look for another way to ask yourself the problem. Try reading it as if both angles TOGETHER equal 35 degrees, then what do we know we should do first?

4. Take this one movement at a time, draw in lightly what happens first and continue to see if it truly is the same shape transformed.

5. Plot the points first, then look at the distances from the points in relation to the x and y axis.

## Chapter 7 Number Sense

### Level 1

1. Can you simplify any of the fractions? Equivalent fractions are the same ratio between the numerator and denominator.

2. It's complicated to divide by a fraction of a whole. We can modify the problem to remove the decimal from the dividend by moving the decimal over in both numbers the same amount of spaces.

3. When a number is larger in the numerator, it is improper. Can you turn it into a mixed number by seeing how many of the denominators you can take out of the numerator?

4. The recipe makes too much, if we want to make a smaller amount what can we do to the recipe? Let's look at how 36 and 12 relate to each other in terms of quantity. 12 is how many times less than 36? Now we know what to do with the recipes.

5. After combining phone times in fraction form, we need to compare that to what it means in terms of minutes. There are 60 minutes in a hour, so we need to see how many minutes are in each $\frac{1}{12}$ of an hour.

### Level 2

1. Remember making factor trees? See what is the largest number they all share.

2. To add, we need common denominators.

3. It's complicated to divide by a fraction of a whole. We can modify the problem to remove the decimal from the dividend by moving the decimal over in both numbers, the same amount of spaces.

4. "KCF" (keep-change-flip) when dividing fractions, then look for equivalents.

5. If one serving is $\frac{2}{3}$ cups and we need 3 cups, how many $\frac{2}{3}$ can go into 3? What operation is this asking? Or we could look at 3 cups as an improper fraction. How the improper fraction compares to the original serving size?

Level 3

1. We still need a common denominator and then we need to follow integer rules.

2. We can try each number in the problem, or we can use what we know about the rules. We need to multiply by a number that is less than a whole. We also need to remember to follow the integers.

3. Remember PEMDAS and integer rules to simplify the expression.

4. For this problem, we need to look at buzz words below and above. Then we are looking at the distance from zero each way. For example, the distance between -2 and 2 is 4.

5. We need a common denominator for the numerator addition problem, and recall if we divide by a fraction, we are multiplying by the reciprocal.

Level 4

1. A perfect square is the product of a number times itself.

2. Count the number of places you need to move over the decimal to get it to the ones place. That will be the exponent to the base ten.

3. 47 is not a perfect square, so we need to find what two perfect squares the number falls in between.

4. PEMDAS

5. Start with solving the problem on your own, then see if you can recognize where your steps differ from the work Miguel produced.

## Chapter 8 Data and Statistics

Level 1

1. The data collected will need to have a visual display for the division of the total surveyed.

2. Each dot represents 1 a game. The numbers represent how many doubles were hit. For example, a dot above the 3 means 3 doubles.

3. The number of days are listed in a 5-degree range of temperature.

4. Look for keywords to see if it could happen or not. Then choose by how often you believe it would happen.

5. Think of what fractions might match up to the percent amount. If the circle was full it would be 100% or 1 whole. So, the denominator must be 100.

Level 2

1. Start by counting what is in the container and what you know is already there. Then you can find out what is missing to make the correct ratio.

2. Place numbers in order from lowest to highest. This way you can clearly see the median and the mode and will better understand the relationship.

3. A 6-sided die has numbers 1,2,3,4,5 and 6. Half the numbers are even. Use that information to figure out the problem.

4. Representative is an unbiased sample of a population.

5. This line plot shows 1 dot per total points scored by the players in the data set. For example, 2 different players scored 4 points, so there are 2 dots above the 4-point mark.

Level 3

1. Likely means it is something that would happen often, more than half the time.

2. The problem gives us the total number of girls in the class and the total number of students in the class. We need to use this to find out the total number of boys in the class and then we will be able to answer the question.

3. We can take what we know about the probability of rolling a 5 one time to solve this problem. If we know how many times it will show up once, we then can find out how many times it will show up for any other number.

4. In order to find this combination, we first need the individual probabilities and then we can use what we know about combinations to get the answer.

5. There are 4 cats for every 7 dogs. How much do you have to multiply to get from 4 cats to 28 cats?

Level 4

1. Look for patterns and trends when deciding what the correlation is on a scatterplot.

2. We need to think about what the range tells us. We know the high and low number and the range is how many choices are in between.

3. A line of best fit is also like a divider line. We want to have the same number of dots on each side of the line, minus the outliers if they are way off.

4. We really need to look at this as 2 choices, not 3. It is either going to be a nickel, or one of the other two coins. How many of those coins are dimes and nickels out of the total 8? Now we have right probability to work with.

5. Remember that when the question speaks of the TOTAL children using sprinkles, it is referring to sprinkles only and sprinkles/chocolate syrup combo. We only know one of those numbers to start.

# Practice Test One

## Level 1 Practice Test

1. There are about 1.2 x $10^7$ people who live in Ohio. Write Ohio's population in standard form.

2. Danny went on a hiking trip for the weekend. On the first day, he hiked the 2 ¾ mile trail. The next morning, he decided to take the 3 ⅛ mile trail. How many total miles did Danny hike on his trip?

3. What is the volume of the figure below in cubic units? Each cube is equal to 1 cubic unit.

4. The picture below shows $\frac{3}{8}$ of Brandyn's leaf collection.

   a. What fraction of Brandyn's leaf collection is 3 leaves?

   b. What fraction of his collection is 21 leaves?

5. What is the value of the expression $1\frac{3}{8} - \frac{3}{4}$ ?

6. Select all the following that would round to 7.

- ❏ 6.93
- ❏ 7.28
- ❏ 6.49
- ❏ 7.88
- ❏ 7.099

7. An area model is shown

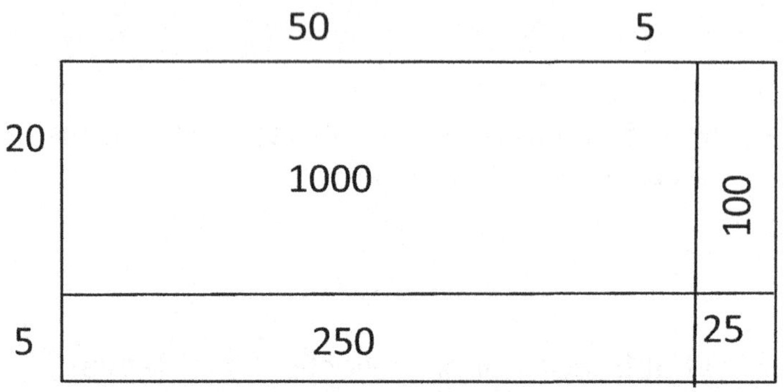


Which problem can be solved using the area model?

A. $1000 \div 25$

B. $1000 \div 250$

C. $1025 \div 55$

D. $1375 \div 25$

8. Simplify the expression

$(8 \times 1{,}000) + (6 \times 100) + (3 \times 10) + (7 \times 1) + (4 \times \frac{1}{10})$

9. The chess club ordered 3 pizzas. Each pizza has 8 slices. If there are 5 members, how many whole slices did each member get? Write as a mixed number, then decide how many whole pieces each member will receive.

10. A local store was charging $3,799 dollars for a laptop. Customers have the option of purchasing insurance for $\frac{1}{10}$ of the cost of the computer. What is the price of the insurance for this item?

11. Select all expressions that have a value less than $\frac{87}{84}$.

A. $\frac{87}{84} \times \frac{101}{100}$

B. $\frac{87}{84} \times \frac{100}{103}$

C. $\frac{87}{84} \times \frac{127}{128}$

D. $\frac{87}{84} \times \frac{200}{1}$

12. Felicia is growing corn in her garden. The chart shows the growth of stalks a, b, and c in inches, for the months of May, June, July, and August. Record the total inches each stalk grew over the four months.

| | May | June | July | August | Total |
|---|---|---|---|---|---|
| A | 10 | 24.86 | 28.3 | 21.22 | |
| B | 8 | 22.6 | 30.48 | 23.38 | |
| C | 12 | 21.83 | 31.4 | 19.17 | |

13. The box needs filled with cubes. Each cube is 1 cubic unit. How many cubes will fill the box? Write your answer in cubic units.

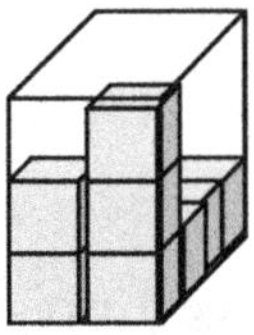

14. Raj has 53 books in his house. Tim has 4/3 as many books as Raj. Explain how you know Tim has the most books.

15. What is the value of 14 divided by 4?

16. Molly buys 8 feet of ribbon to decorate her room. She uses ¼ of the ribbon around her mirror and then uses ⅓ of what is left of the ribbon to tie back her curtains. How many feet of ribbon are left?

17. Miko has 1,220 songs stored on her phone. Toni only has 1/10 as many songs. How many songs does Toni have on her phone?

18. What is the total volume of the figures below?

# Level 2 Practice Test

1. What is the distance between (-3, 8) and (2, 8) on a coordinate grid?

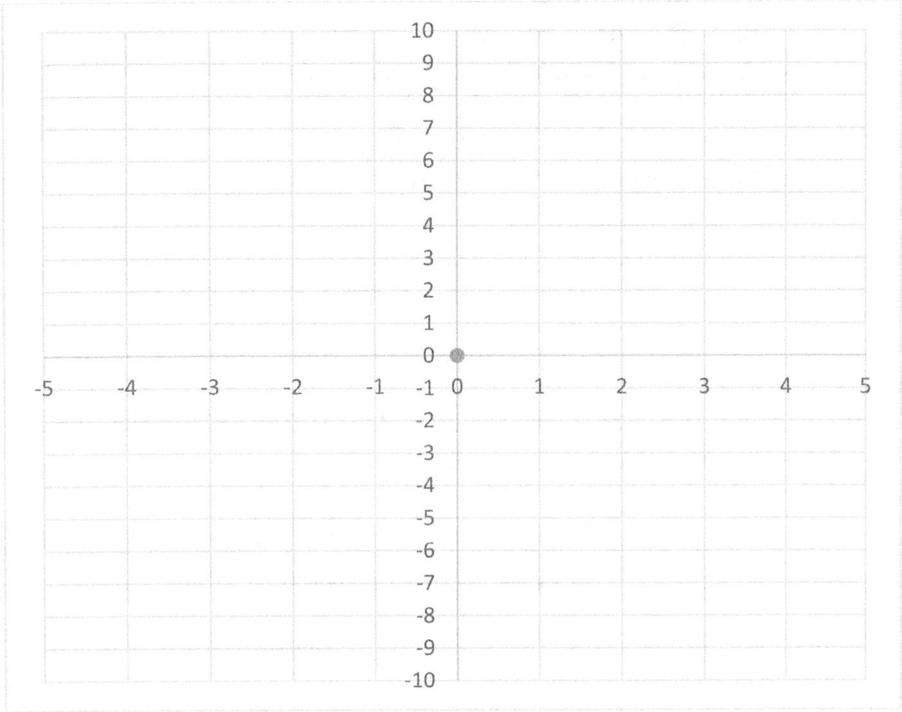

2. Select three expressions that are equivalent to 4b + 8c.

   A. b + 2c
   B. 2(2b + 4c)
   C. 3b + b + 4c + 4c
   D. ¼ (b + 2c)
   E. 4(b + 2c)
   F. ⅛ (4b + 8c)

3. A data set is displayed in the dot plot shown.

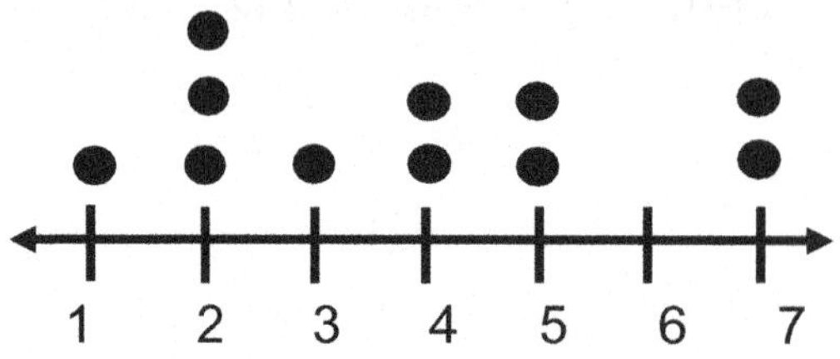


What is the mean of the data set?

4. Label the following equations with the appropriate math terms from the list (use all terms)

$(2x)(c) = 24$

* product
* term
* Coefficient

5. The students took a field trip to the zoo. The equation models the total cost of the trip, where x is the number of students who attended.

$$12x = 252$$

How many students went to the zoo?

6. There are 433 students in Johnsonville High School. If the number of boys is 85 more than the number of girls in the school, how many girls are in the school?

7. The area of the kitchen floor is 40 $yd^2$. Joe will cover the floor with square tiles that each have an area of 1 square foot ($ft^2$). How many tiles does Joe need to completely cover the floor?

8. Select the three statements that represent a unit rate.

   A. Apples cost $2 per pound
   B. A teacher divides the class so she has 7 students in each group
   C. Shawn buys 3 shoes for $90
   D. A recipe calls for 3 tbls. Salt for every gallon of water
   E. Julio can drive 230 miles in 4 hours

9. Select the box to match each expression on the left to the equivalent expression.

| | 24 + 54 | 8 + 12 | 8 + 32 | 24 + 9 |
|---|---|---|---|---|
| 3(8 + 3) | | | | |
| 6(4 + 9) | | | | |
| 4(2 + 8) | | | | |

10. Janice makes a box in the shape of a rectangle. A net of the box is shown. What is the surface area of the box?

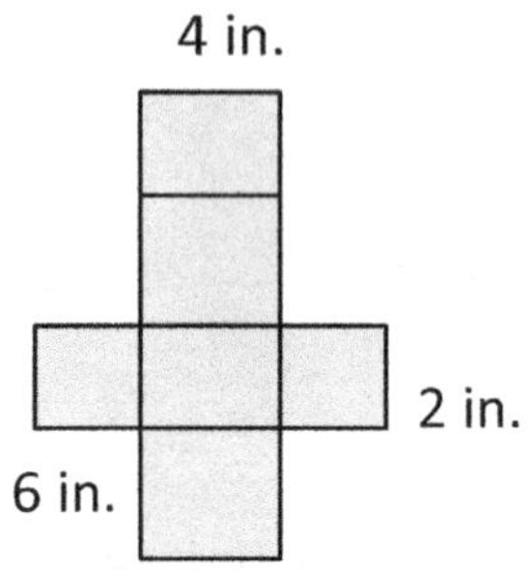


11. An expression is given.

*4h - 9*

Which phrase describes the expression?

A. The product of 9 subtracted from 4 and a number h
B. The difference of 4 plus a number h and 9
C. 9 subtracted from the product of 4 and a number h
D. The difference of 4 plus a number h and 9

12. Cristina makes 3 containers of hot sauce that fill a total of 90 bottles. Each container of hot sauce costs $30 to make. What is the cost per bottle for the hot sauce?

13. What is 62% of 28?

14. Kendra wants to exchange her gem collection at a rate of $1 for 0.75 lbs. How much money will she receive for 3 lbs?

15. Aniya wants to ask her class a statistical question. Select all questions that are statistical.
    A. What day do you prefer to take a test on?
    B. Who is your teacher?
    C. How many class periods do you have before lunch?
    D. Do you prefer a seat toward the front or the back of the class?
    E. How many days are in the school year?

16. Simplify the expression $\frac{1}{4}$ (8 x 2.6) .

17. Annabelle went to pick strawberries in the field. On Monday, she picked 6 $\frac{7}{8}$ lbs., on Thursday she collected 4 $\frac{1}{2}$ lbs. How many more pounds did she pick on Monday?

## LEVEL 3 PRACTICE TEST

1. Which two data sets have the most overlap?

A. 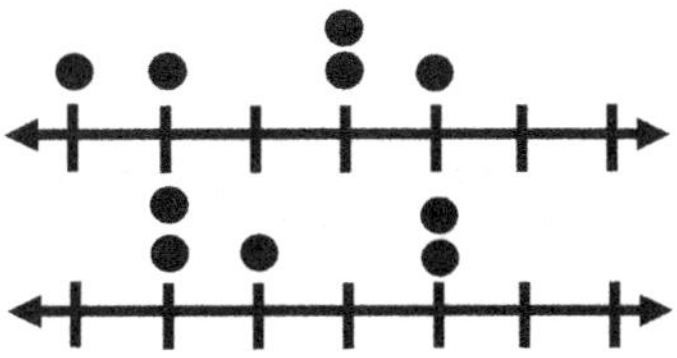

C. 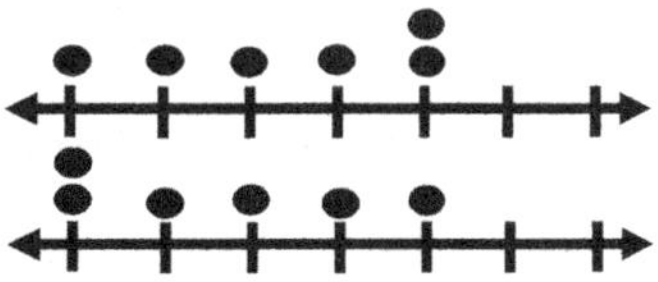

B. 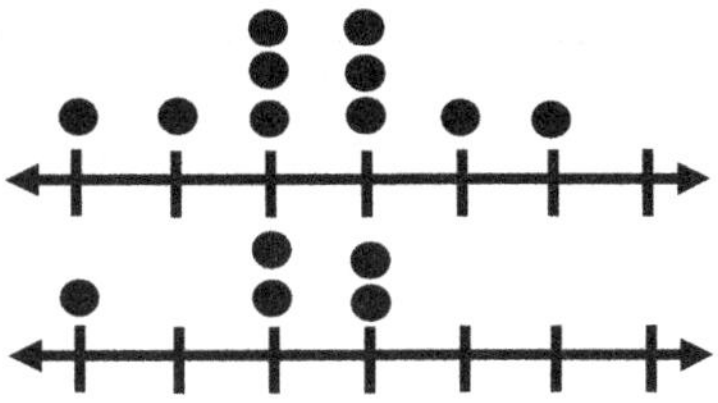

D. 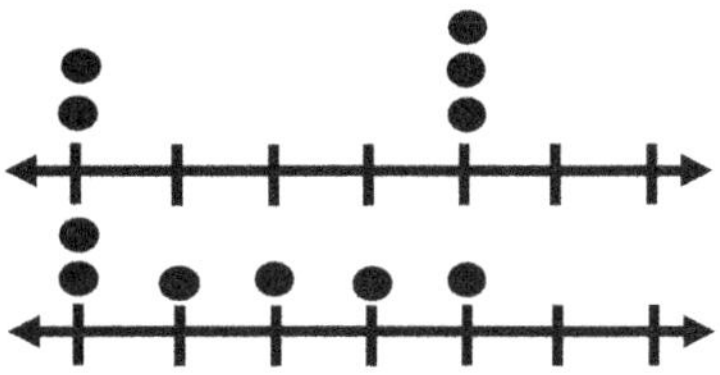

2. A triangle has a side length of ½ inch and a side length of 4 inches. What could be a possible length for the third side of the triangle?

3. The relationship between the number of cups of sugar, s, and the number of cups of water, w, used in a recipe is described by the following equation shown.

$$w = 3\ s$$

How many cups of water are needed for each cup of sugar?

4. The graph shows the relationship between the water level in a tank as it is filled over time.

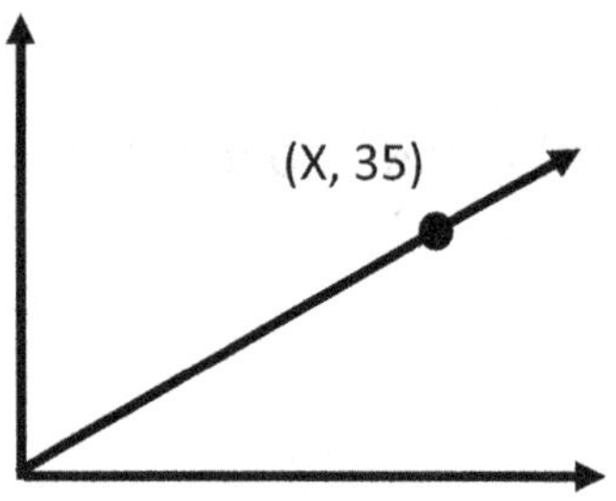


What does the point (x, 35) represent in the situation?

A. The tank fills for x seconds
B. The tanks fills 35 gallons per second
C. The tank fills y gallons per second
D. The tank fills 35 gallons every x seconds

5. Ms. Kraner arranges her preschool students into three groups of equal size, labeled red, blue, and green. Each day she selects one group at random to clean up the stations at the end of the day. The data for 6 days are shown.

| Day | Group |
|---|---|
| 1 | Yellow |
| 2 | Yellow |
| 3 | Red |
| 4 | Blue |
| 5 | Red |
| 6 | Yellow |

Which group was chosen at the expected frequency?

6. A three dimensional figure is shown. What shape will result in the cross section display if the figure is cut parallel to the base?

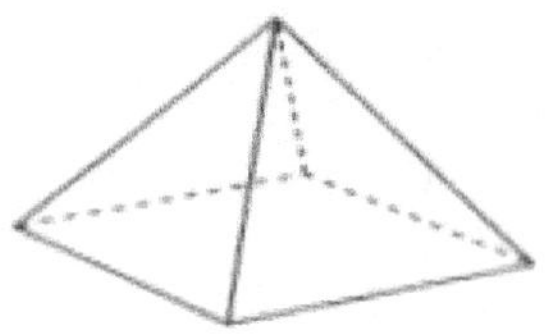

7. An inequality is given, where k is a positive number.

   $k \cdot \frac{m}{k} \geq 0$

   Which inequality must also be true?

   A. $m \geq 0$
   B. $m \leq 0$
   C. $m > -k$
   D. $m \leq -k$

8. What is the volume, in cubic centimeters of the triangular prism?

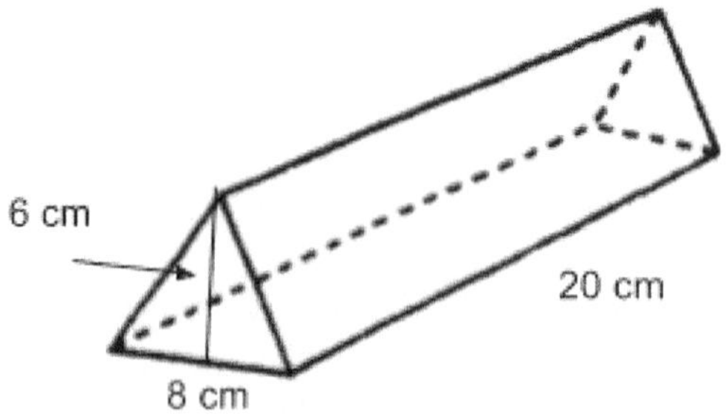

9. A dripping faucet fills a bucket at a rate of 1/20 of a gallon in ⅕ of an hour. How much water, in gallons, will be in the bucket after 1 hour?

10. A newspaper wants to determine how often it delivers to three different areas of a city. The table shows the areas from a random sample of 120 deliveries.

| Area of the city | # of deliveries |
|---|---|
| north | 36 |
| central | 55 |
| South | 29 |

Based on this information, if the driver makes 400 deliveries, how many deliveries will be to the north end of town?

11. Two options for going to a fitness class at the gym are available.

- Without membership: \$8.50 per class
- With membership: unlimited classes for \$35 per month

Brittany plans to attend 1 class a week for a year. Calculate the cost for Brittany under each option, then justify whether Brittany should pay per class or buy the membership.

12. William has a coupon for an additional 10% off the sale price of a single item in the store. He uses the coupon to buy a bike that was tagged with the sale price of 25% off the original price. He pays \$60.75 and there is no sales tax. What was the original price of the bike?

13. Which of the following are equivalent expressions to $-6(x - 10) + 5x$

A. $-6x - 60 + 5x$
B. $-6x + 60 + 5x$
C. $-6x - 16 + 5x$
D. $-6x + 4 + 5x$
E. $11x + 60$
F. $-x + 60$

14. A clothing store is having a sale for 20% off everything. Benjamin states the sale can be represented by the expression $x - .20x$. Allison thinks her expression, $.80x$, is correct.

Prove why both actually will represent the same outcome and explain why.

15. Simplify the expression $\frac{-3}{7} + \frac{8}{9}$ .

16. The water level for Lake Anna is 2 inches above sea level at the end of the week. What are possible levels for Monday and Thursday?

| Day | Level |
|---|---|
| Sunday | 5 |
| Monday | |
| Tuesday | -7 |
| Wednesday | 1 |
| Thursday | |
| Friday | -1 |
| Saturday | 3 |

17. A jar contains red marbles, blue marbles, and green marbles. There are more red marbles than green marbles. The probability of selecting a red marble from the jar is ¼. What is a possible probability that a randomly selected marker from the jar is green?

18. An equation is shown where $b<0$ and $c>0$.

$$a - b = c$$

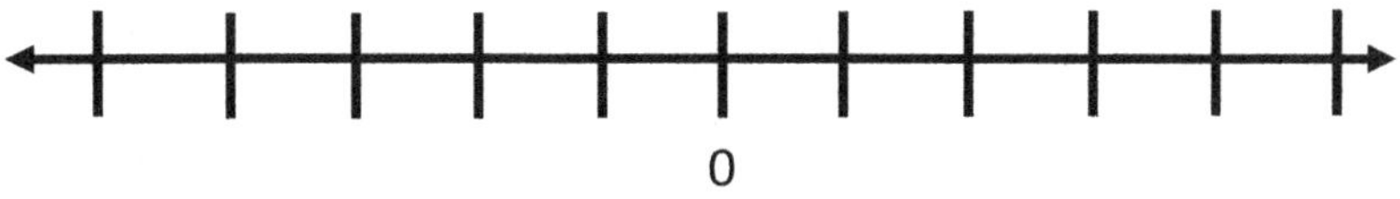


Place possible points for a, b, and c on the number line.

# Level 4 Practice Test

1. A scatterplot is shown.

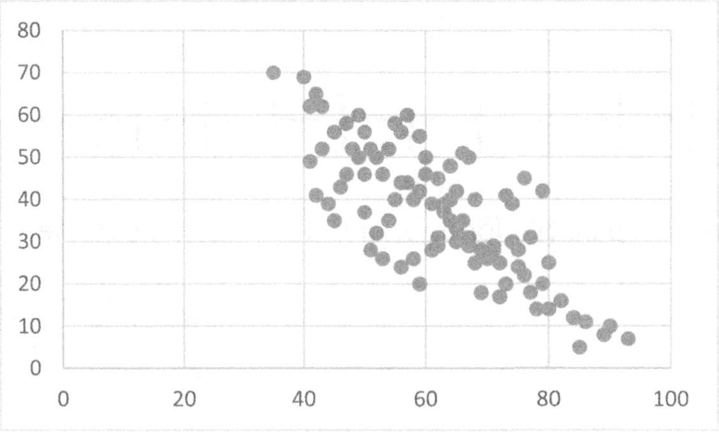


Which statement about the scatterplot is true?

A. There is no correlation
B. There is a negative, linear correlation between x and y
C. There is a positive, linear correlation between x and y
D. There is a nonlinear relationship between x and y

2. Select all equations that represent a nonlinear function.

A. Y = 6x - 14
B. Y = $5x^2$+ 4x -18
C. Y = x(2x- 10)
D. Y = 7x + 8 + x
E. Y = ½ x - 16

3. Simplify the following expression in scientific notation.

$6.4 \times 10^5 + 2.3 \times 10^6$

4. Identify if each equation has no solution, one solution, or many solutions.

| | No solution | One solution | Many solutions |
|---|---|---|---|
| -6x + 2 = -6x + 2 | | | |
| -8x + 7 = 8x + 7 | | | |
| -9x + 8 = -8x + 9 | | | |

5. What is the distance between (3,6) and (9,11)? Use the coordinate plane and Pythagorean theorem.

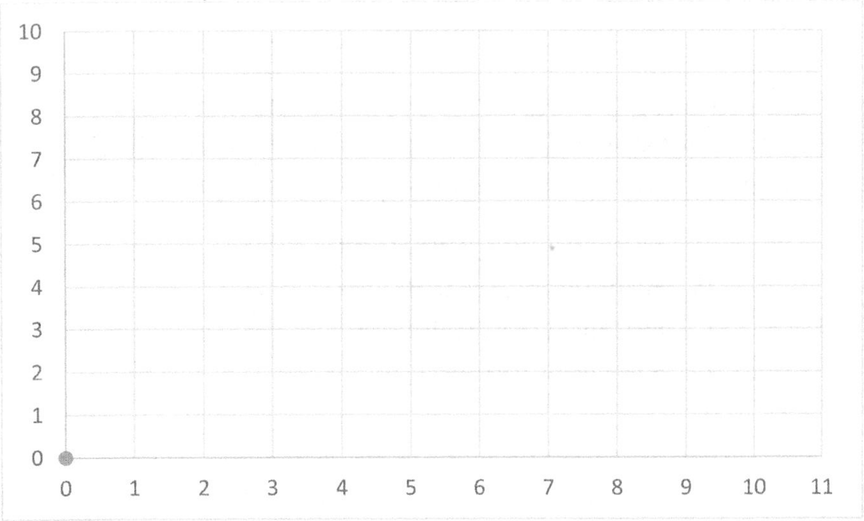

6. Select all expressions that are equivalent to

$$\frac{(4^2)^4 \cdot 4^8}{4^6}$$

A. $4^{10}$

B. $4^{14}$

C. $4^8$

D. $\frac{4^{14}}{4^6}$

E. $\frac{4^{16}}{4^6}$

7. Two quadrilaterals are shown

Garrett uses two transformations to show that quadrilateral ABCD is similar to quadrilateral JKLM. What pair of transformations did Garrett use?

A. Rotation and translation
B. Reflection and translation
C. Rotation and reflection
D. Rotation and dilation

8. A triangle XYZ has vertices X (0,2), Y (3,6), and Z (7,4). Triangle X'Y'Z' is created by reflecting triangle XYZ over the x-axis.
   What is the coordinate of vertex Z'?

9. An inequality is given, where *g* is a whole number.

   $$\frac{1}{g} > \frac{1}{\sqrt{23}}$$

   Create an inequality to show all possible values of *g*. Explain your answer.

10. Two canisters are shown. What is the combined volume, in cubic inches, of the canisters? Round your answer to the nearest hundredth.

11. Which graph represents a function that is decreasing at a constant rate?

A.

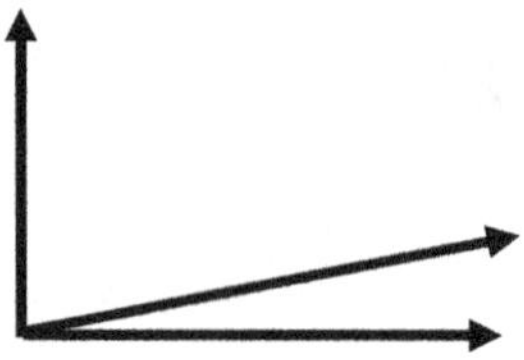

B.

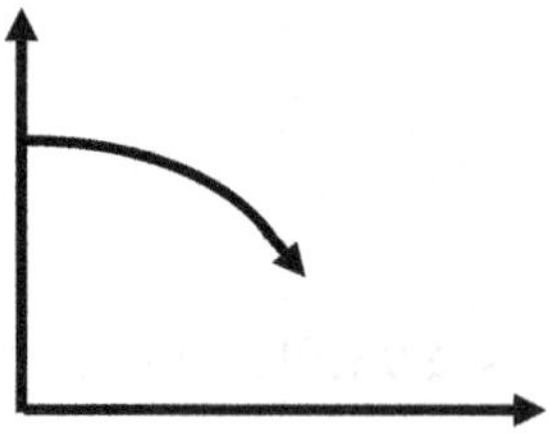

C.

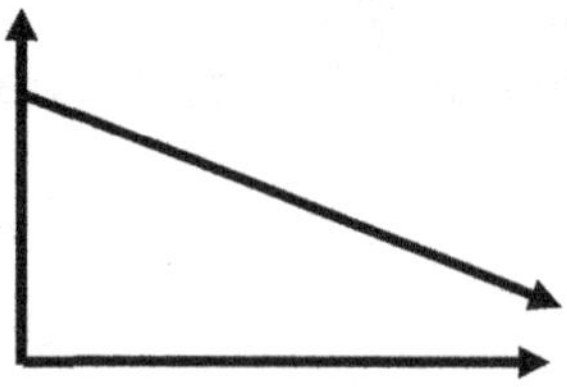

D.

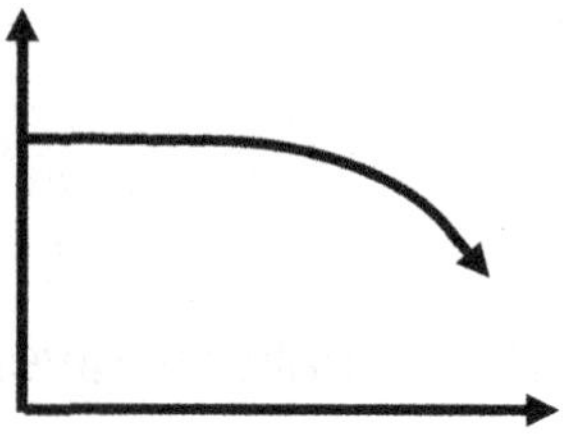

12. The speed of an object is calculated using the following formula

$$\text{Speed} = \frac{distance}{time}$$

A shooting star traveled $8.6 * 10^5$ miles in about 1.6 x $10^4$ seconds. What is the speed of the star?

13. Braylen and Dionte collect baseball cards. The given equations model the number of cards b, that Braylen has and the number of cards d, that Dionte has.

$$b + d = 87$$

$$d = b - 15$$

How many cards does Dionte have?

14. Isosceles trapezoid HIJK is shown. It is reflected to create trapezoid LMNO.

Which statement is true?

A. $H \simeq M$
B. LM || NO
C. MN || LO
D. $J \simeq L$

15. What is the value of y in terms of x if *a* is parallel to *b*?

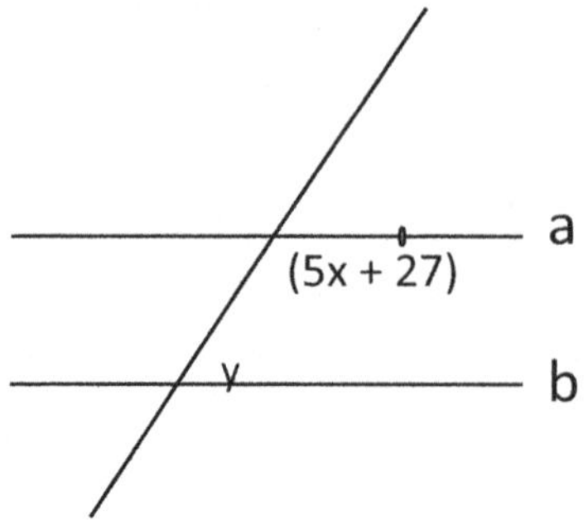


16. An equation is shown.

$$7^a \cdot 7^b = \frac{1}{7}$$

What are 2 possible values for both a and b?

17. The value of $6 \times 10^{12}$ is __________ times greater than the value of $3 \times 10^9$.

What number correctly completes this sentence?

18. A table of values for Function A and a graph of function B are shown.

Function A

| X | f(X) |
|---|---|
| 0 | 1 |
| 2 | 5 |
| 3 | 7 |
| 6 | 13 |
| 10 | 21 |

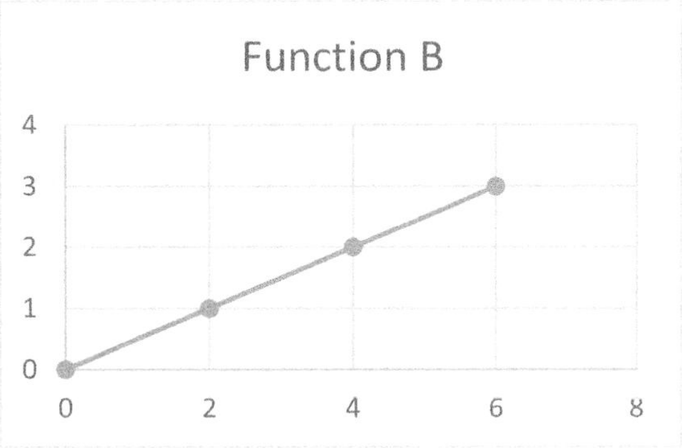


Create an equation for a third linear function where the ratio of change is between the rates of change for function A and B.

# Practice Test 2

## Level 1 Practice Test 2

1. There are about 1.57 x $10^8$women who live in the United States. Write the population of women in standard form.

2. Claire went on a hiking trip for the weekend. On the first day, she hiked the 1 ¼ mile trail. The next morning she decided to take the 2 ⅞ mile trail. How many total miles did Claire hike on her trip?

3. What is the volume of the figure below in cubic units? Each cube is equal to 1 cubic unit.

4. The picture below shows $\frac{1}{6}$ of Austin's rock collection.

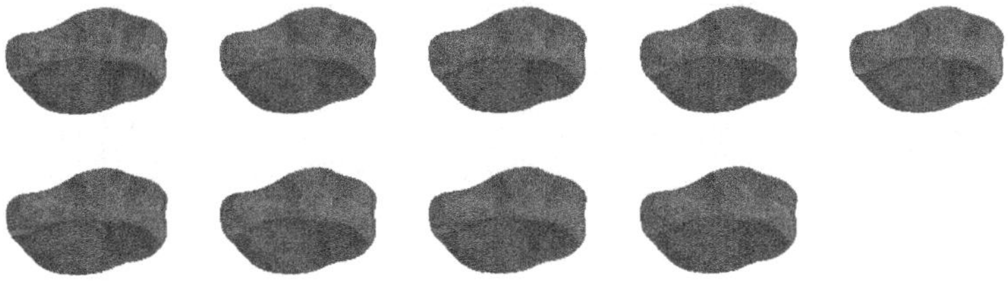

   a. What fraction of Austin's rock collection is 4 rocks?

   b. What fraction of his collection is 40 rocks?

5. What is the value of the expression $2\frac{3}{4} - \frac{5}{8}$ ?

6. Select all the following that would round to 9.

   - ❑ 9.92
   - ❑ 9.27
   - ❑ 8.41
   - ❑ 8.088
   - ❑ 8.91

7. An area model is shown

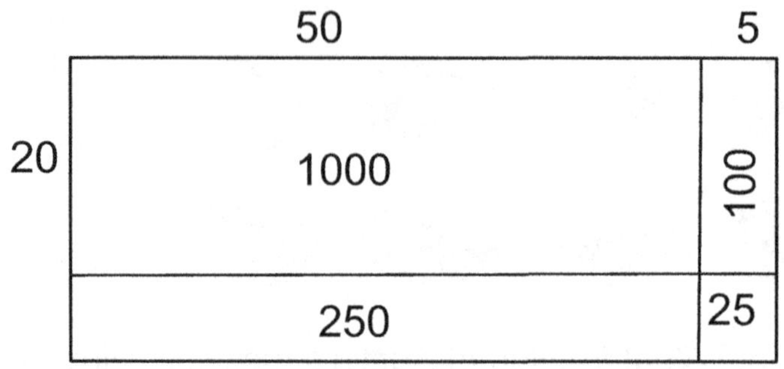


Which problem can be solved using the area model?

A. 1000 ÷ 25

B. 1375 ÷ 55

C. 1025 ÷ 55

D. 1375 ÷ 100

8. Simplify the expression

$$(6 \times 1,000) + (2 \times 100) + (5 \times 10) + (7 \times 1) + (8 \times \frac{1}{10})$$

9. The archery club ordered 4 pizzas. Each pizza has 8 slices. If there are 7 members, how many whole slices did each member get? Write as a mixed number, then decide how many whole pieces each member will receive.

10. A local store was charging $897 dollars for a flat screen television. Customers have the option of purchasing insurance for $\frac{1}{10}$ of the cost of the computer. What is the price of the insurance for this item?

11. Select all expressions that have a value less than $\frac{75}{72}$.

A. $\frac{75}{72}$x $\frac{101}{100}$

B. $\frac{75}{72}$x $\frac{100}{101}$

C. $\frac{75}{72}$x $\frac{127}{128}$

D. $\frac{75}{72}$x $\frac{100}{1}$

12. Carlos is growing corn in his garden. The chart shows the growth of stalks a, b, and c in inches, for the months of May, June, July, and August. Record the total inches each stalk grew over the four months.

| | May | June | July | August | total |
|---|---|---|---|---|---|
| A | 11 | 21.86 | 28.1 | 21.29 | |
| B | 9 | 22.4 | 31.8 | 22.98 | |
| C | 13 | 20.63 | 30.14 | 19.07 | |

13. The box needs filled with cubes. Each cube is 1 cubic unit.

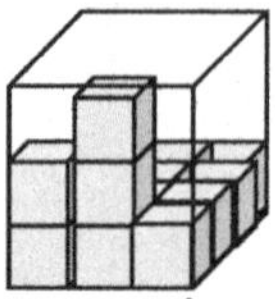

How many cubes will fill the box? Write your answer in cubic units.

14. Sal has 62 coins in his collection. Jacob has 5/4 as many coins as Sal. Explain how you know Jacob has the most books.

15. What is the value of 18 divided by 5?

16. Miranda buys 12 feet of ribbon to make bows. She uses ¼ of the ribbon for the first bow and then uses ⅓ of what is left of the ribbon to make a second bow. How many feet of ribbon are left?

17. Mitch has 1,090 songs stored on his phone. Triston only has 1/10 as many songs. How many songs does Triston have on his phone?

18. What is the total volume of the figures below?

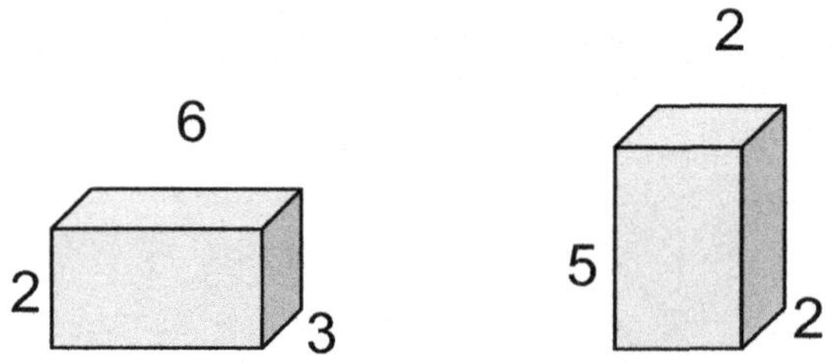

# Level 2 Practice Test 2

Level 2 practice test

1. What is the distance between (-1, 6) and (4, 6) on a coordinate grid?

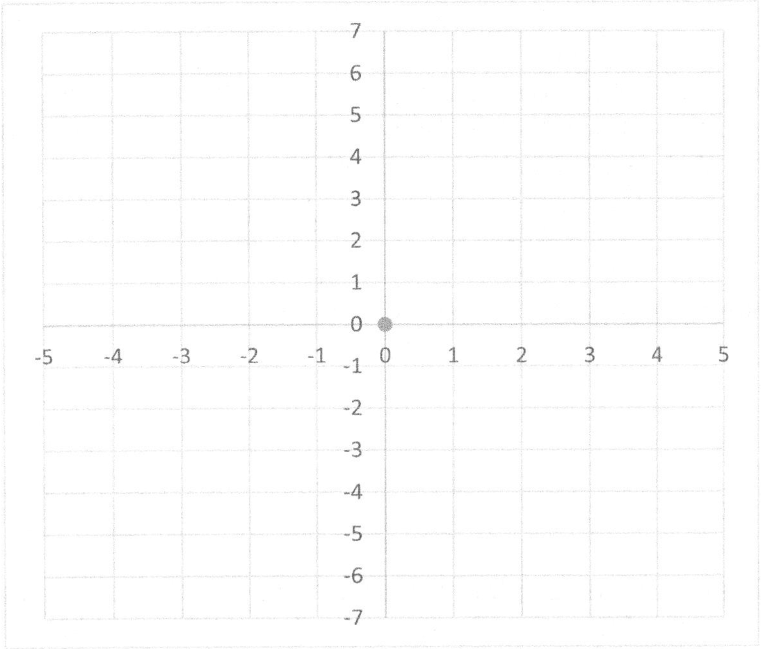

2. Select three expressions that are equivalent to 6w + 12c.
   A. w + 2c
   B. 2(3w + 4c)
   C. 3w + 3w + 6c + 6c
   D. ¼ (w+ 3c)
   E. 3(2w + 4c)
   F. ⅙ (4w + 8c)

3. A data set is displayed in the line plot shown.

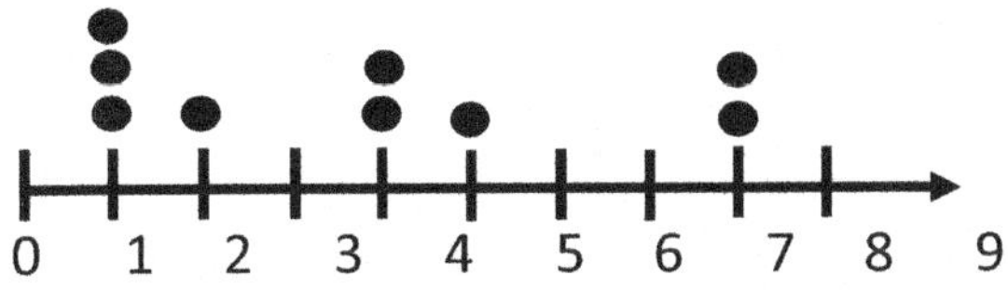

   What is the mean of the data set?

4. Label the following equations with the appropriate math terms from the list (use all terms)

(4b)(h) = 38

* product
* term
* Coefficient

5. The students took a field trip to the aquarium. The equation models the total cost of the trip, where x is the number of students who attended.

$8x = 472$

How many students went to the aquarium?

6. There are 381 students in Stormy Middle School. If the number of boys is 47 more than the number of girls in the school, how many girls are in the school?

7. The area of the kitchen floor is 38 $yd^2$. Joe will cover the floor with square tiles that each have an area of 1 square foot ($ft^2$). How many tiles does Joe need to completely cover the floor?

8. Select the three statements that represent a unit rate.

   A. Bananas cost $0.49 per pound
   B. A teacher divides the class so she has 5 students per group
   C. Elaine buys 3 shirts for $60
   D. A recipe calls for 1 tbls. salt for every cup of water
   E. Julio can drive 46 miles each hour

9. Select the box to match each expression on the left to the equivalent expression.

| | 24 + 54 | 8 + 12 | 8 + 32 | 24 + 9 |
|---|---|---|---|---|
| 3(8 + 3) | | | | |
| 6(4 + 9) | | | | |
| 4(2 + 8) | | | | |

10. Janice makes a box in the shape of a rectangle. A net of the box is shown. What is the surface area of the box?

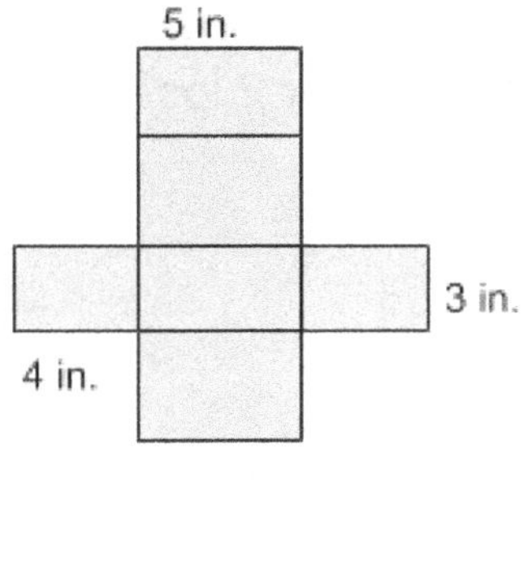

11. An expression is given.

$2m + 7$

Which phrase describes the expression?

A. The product of 7 and 4 plus a number *m*
B. The sum of 7 plus a number *m* times 2
C. 7 subtracted from the product of 2 and a number *m*
D. The difference of 7 plus a number *m* and 2

12. Calvin makes 4 containers of hot sauce that fill a total of 120 bottles. Each container of hot sauce costs $20 to make. What is the cost per bottle for the hot sauce?

13. What is 62% of 28?

14. Kara wants to buy strawberries at a price of $1 for 0.4 lbs. How much money will she pay for 2 lbs?

15. Adam wants to ask his community a statistical question. Select all questions that are statistical.

   A. What day do you prefer to go out to eat on?
   B. Who is our major?
   C. What street do you live on?
   D. Do you prefer walking or driving around town?
   E. How many people live in the town this year?

16. Simplify the expression ½ (12 x 1.3).

17. Alexandra went to pick blueberries in the field. On Monday she picked 7⅚ lbs., on Thursday she collected 5 ½ lbs. How many more pounds did she pick on Monday?

## Level 3 Practice Test 2

1. Which two data sets have the most overlap?

A.

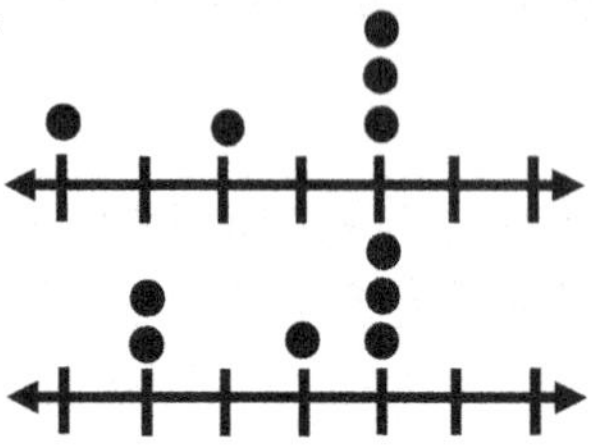

C.

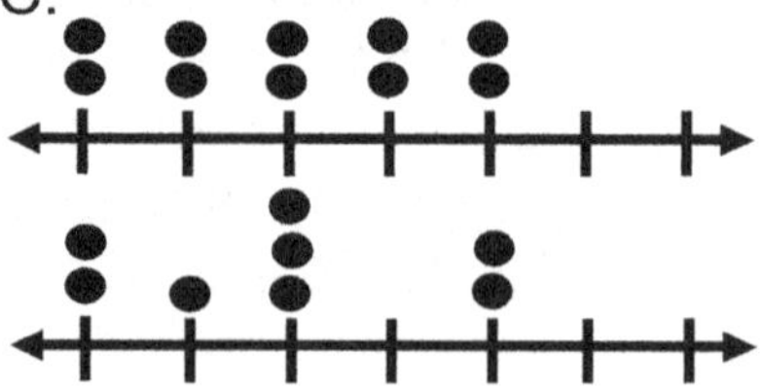

B.

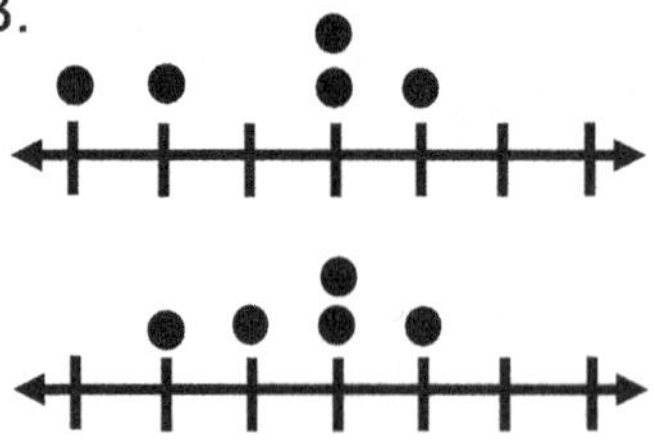

D.

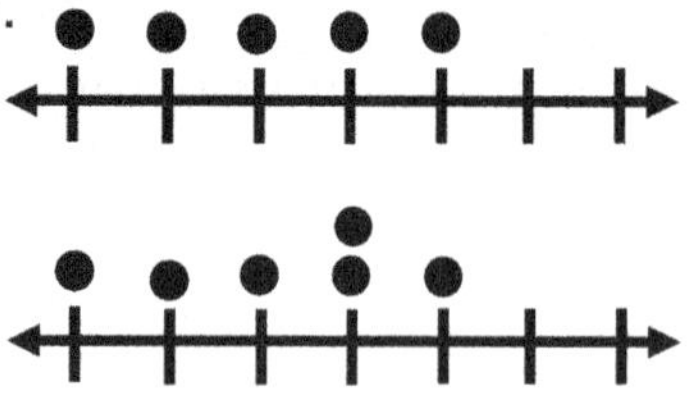

2. A triangle has a side length of 2 inches and a side length of 6 ½ inches. What is a possible length for the third side of the triangle?

3. The relationship between the number of cups of flour, f, and the number of cups of milk, m, used in a recipe is described by the following equation shown.

$$m = 2f$$

How much milk is needed for each cup of flour?

4. The graph shows the relationship between the water level in a tank as it is filled over time.

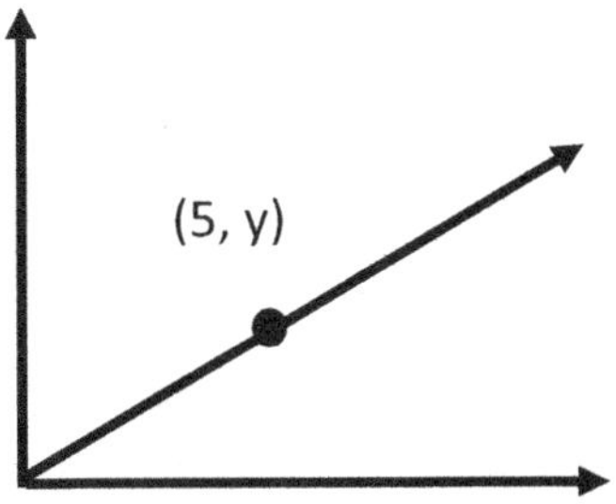


What does the point (5, y) represent in the situation?

A. The tank fills y gallons every 5 seconds
B. The tank fills for y seconds
C. The tank fills 5 gallons per second
D. The tank fills y gallons per second

5. Mr. Paine divided his students into three groups of equal size, labeled red, blue, and green. Each day he selects one group at random to clean up the stations at the end of the day. The data for 6 days are shown.

| Day | Group |
|---|---|
| 1 | Yellow |
| 2 | Blue |
| 3 | Red |
| 4 | Blue |
| 5 | Red |
| 6 | Red |

Which group was chosen at the expected frequency?

6. A three-dimensional figure is shown.

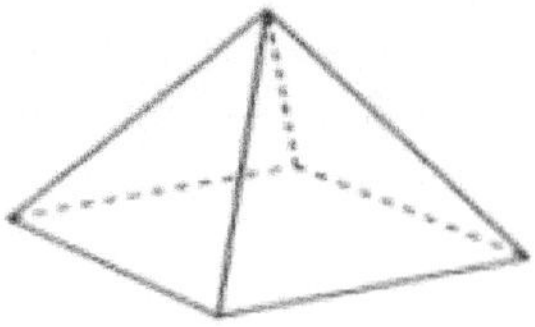

What shape will result in the cross section display if the figure is cut perpendicular to the base, from the apex?

7. An inequality is given, where k is a negative number.

$$k \cdot \frac{m}{k} \geq 0$$

Which inequality must also be true?

A. $m > -k$
B. $m \leq 0$
C. $m \geq 0$
D. $m = k$

8. What is the volume, in cubic centimeters of the triangular prism?

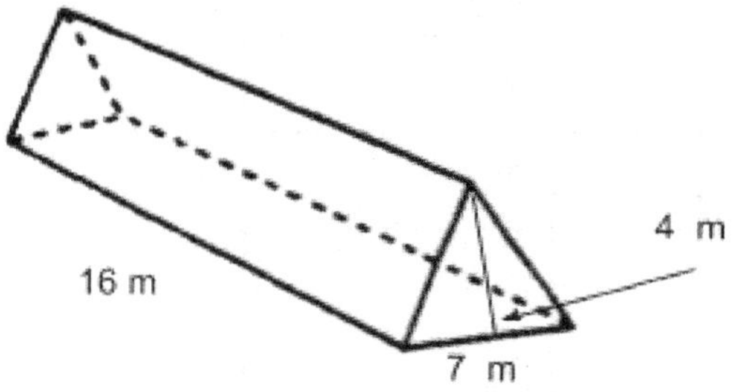

9. A dripping faucet fills a bucket at a rate of 1/10 of a gallon in ⅛ of an hour. How much water, in gallons, will be in the bucket after 1 hour?

10. A florist wants to determine how often it delivers to three different areas of a city. The table shows the areas from a random sample of 40 deliveries.

| Area of the city | # of deliveries |
|---|---|
| North | 12 |
| Central | 19 |
| South | 9 |

Based on this information, if the driver makes 200 deliveries, how many deliveries will be to the south end of town? Round to the nearest whole number.

11. Two options for swimming at the rec center are available.

- Without membership: $6.25 per day
- With membership: unlimited days for $61 per month

Blake plans to swim three times a week for the 3 months (13 weeks) of summer. Calculate the cost for Blake under each option, then justify whether Blake should pay per day or buy the membership.

12. Alicia has a coupon for an additional 20% off the sale price of a single item in the store. She uses the coupon to buy a watch that was tagged with the sale price of 20% off the original price. She pays $36.80 and there is no sales tax. What was the original price of the bike?

13. Which of the following are equivalent expressions to ½ (x - 8) + 1.5x ?

a. ½ x - 8 + 1 ½ x
b. ½ x - 4 + 1.5x
c. 2x - 4
d. 2x +4 + 1.5x
e. ½ x + 4 + 1.5x
f. ½ x - 4 + 1 ½ x

14. A clothing store is saving a sale for 30% off everything. Brianna states the sale can be represented by the expression

x - .30x

Alexia thinks her expression is correct

.70x

Prove why both actually will represent the same outcome and explain why.

15. Simplify the expression $\frac{-3}{8} + \frac{5}{12}$

16. The water level for Lake Anna for the week is 5 inches above sea level. What are possible levels for Monday and Thursday?

| Day | level |
|---|---|
| Sunday | 6 |
| Monday | |
| Tuesday | -4 |
| Wednesday | -1 |
| Thursday | |
| Friday | 1 |
| Saturday | -3 |

17. A jar contains red marbles, blue marbles, and green marbles. There are more blue marbles than green marbles. The probability of selecting a blue marble from the jar is ¼. What is a possible probability that a randomly selected marble from the jar is red?

18. An equation is shown where $b > 0$ and $c < 0$. What is a possible placement for a and b?

$$a - b = c$$

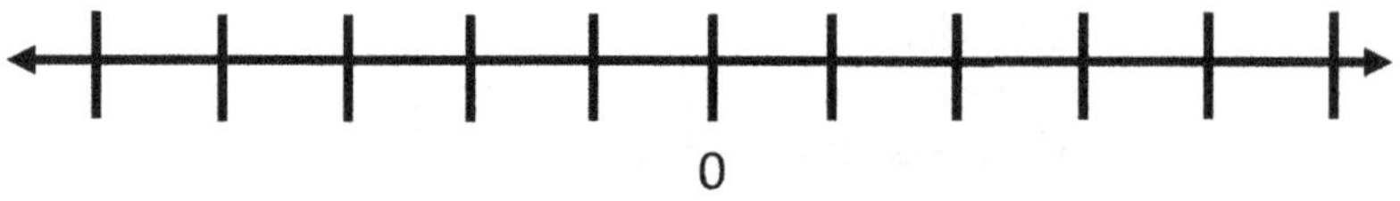


Place possible points for a, b, and c on the number line.

# Level 4 Practice Test 2

1. A scatterplot is shown.

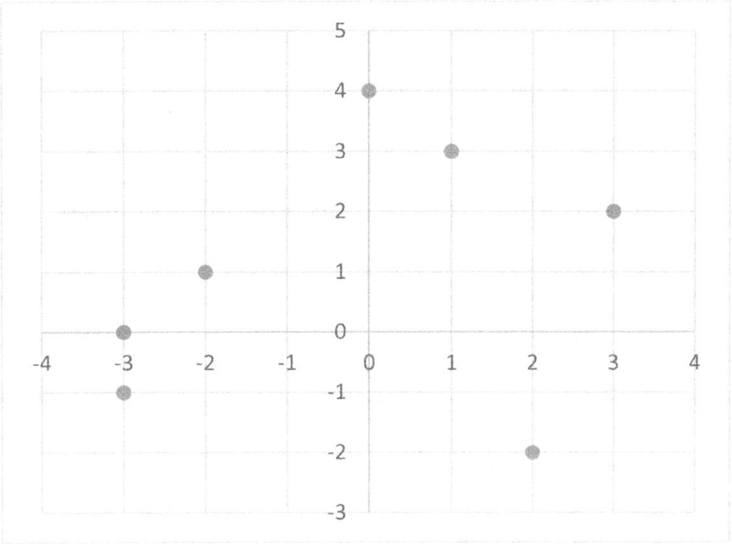


Which statement about the scatterplot is true?

A. There is no correlation
B. There is a negative, linear correlation between x and y
C. There is a positive, linear correlation between x and y
D. There is a nonlinear relationship between x and y

2. Select all equations that represent a nonlinear function.

A. $Y = 3x^2 + 3x - 3$
B. $Y = 6x - 11$
C. $Y = x(2x - 10)$
D. $Y = 14x + 2 + x$
E. $Y = \frac{1}{2} x^2 - 28$

3. Simplify the following expression in scientific notation.

$$1.86 \times 10^{8} - 2.3 \times 10^{7}$$

4. Identify if each equation has no solution, one solution, or many solutions.

| | No solution | One solution | Many solutions |
|---|---|---|---|
| 6x + 11 = -6x + 11 | | | |
| -8x + 7 = -8x + 7 | | | |
| -3x + 4 = -4x + 3 | | | |

5. What is the distance between (2,5) and (6, 7)? Use the coordinate plane.

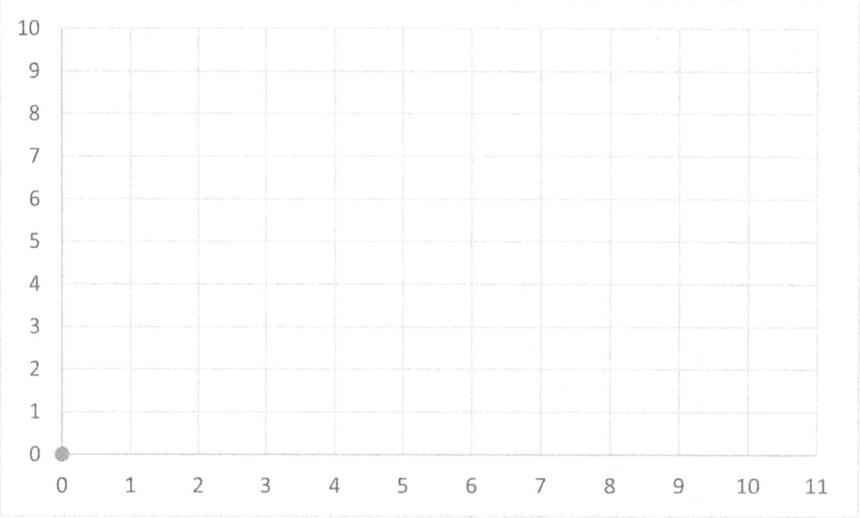

6. Select all expressions that are equivalent to

$$\frac{(6^2)^3 * 6^6}{6^4}$$

A. $6^{12}$

B. $6^3$

C. $6^9$

D. $\frac{6^{12}}{6^4}$

E. $\frac{6^{11}}{6^4}$

7. Two quadrilaterals are shown

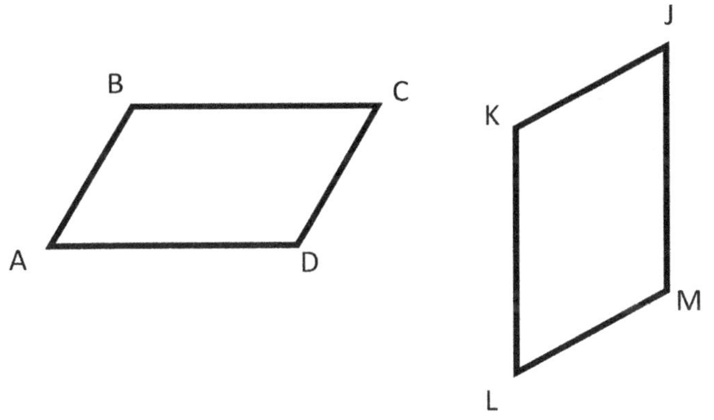


Garrett uses two transformations to show that quadrilateral ABCD is similar to quadrilateral JKLM. What pair of transformations did Garrett use?

A. Dilation and translation
B. Reflection and translation
C. Rotation and reflection
D. Rotation and dilation

8. A triangle XYZ has vertices X (1,3), Y (3,8), and Z (6,4). Triangle X'Y'Z' is created by reflecting triangle XYZ over the y-axis.

What is the coordinate of vertex X'?

9. An inequality is given, where *d* is a whole number.

$$\frac{1}{d} > \frac{1}{\sqrt{47}}$$

Create an inequality to show all possible values of *d*. Explain your answer.

10. Two canisters are shown. What is the combined volume, in cubic inches, of the canisters? Round your answer to the nearest hundredth.

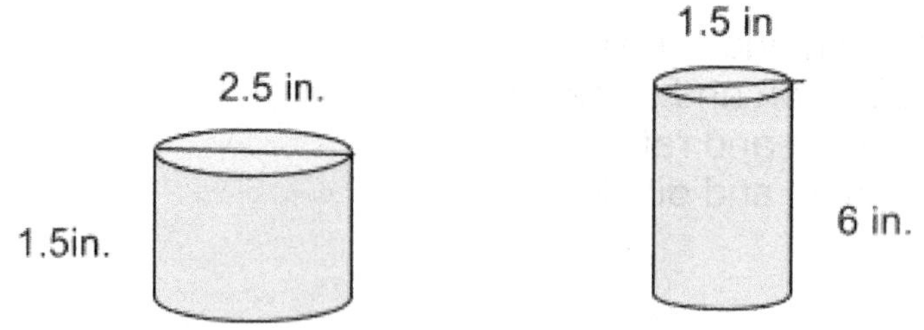

11. Which graph represents a function that is increasing at a constant rate?

A.

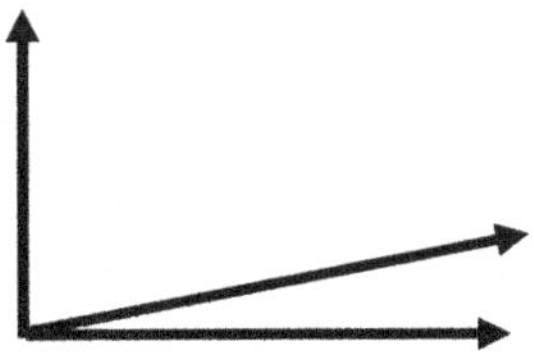

C.

B.

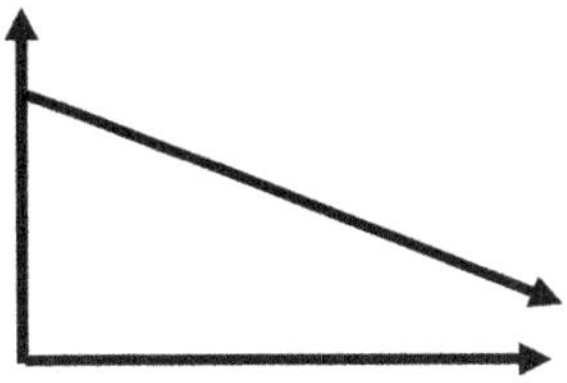

D.

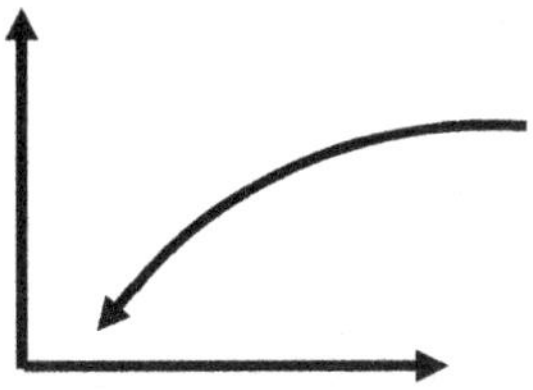

12. The speed of an object is calculated using the following formula

$$\text{Speed} = \frac{distance}{time}$$

A rocket traveled $1.76 \times 10^4$ miles in about $4.4 \times 10^3$ seconds. What is the speed of the star?

13. Bryan and Derrick collect baseball cards. The given equations model the number of cards b, that Bryan has and the number of cards d, that Derrick has.

$$b + d = 86$$
$$d = b - 18$$

How many cards does Derrick have?

14. Isosceles trapezoid HIJK is shown. It is reflected to create trapezoid LMNO.

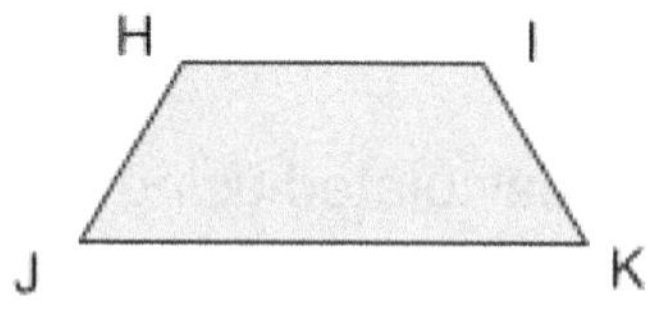


Which statement is true?

A. angle H ≅ angle M
B. HJ || IK
C. MN || LO
D. angle K ≅ angle O

15. What is the value of y in terms of x if *a* is parallel to *b*?

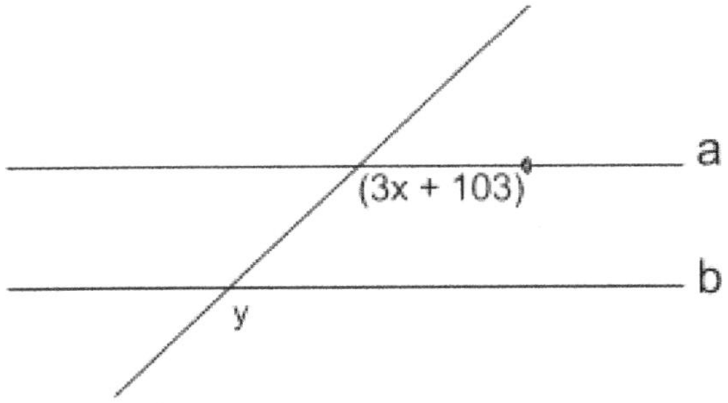


16. An equation is shown

$$4^a \; 4^b = \frac{1}{4}$$

What are 2 possible values for both a and b?

17. The value of $1.8 \times 10^{10}$ is __________ times greater than the value of $6 \times 10^7$.
What number correctly completes this sentence?

18. A table of values for function A and a graph of function B are shown.

| F | F(x) |
|---|---|
| 2 | 1 |
| 4 | 2 |
| 8 | 4 |
| 10 | 5 |
| 16 | 8 |

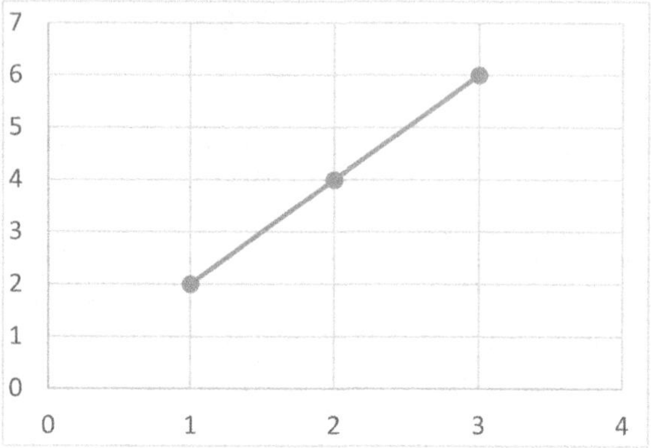


Create an equation for a third linear function where the rate of change is between the rates of change for functions A and B.

# Practice Test Answer Key

# Practice Test 1

## Level 1

1. 12, 000, 000 people
2. 5 ⅞ miles
3. 18 cubic units
4. a. ⅛ b. ⅞
5. ⅝
6. 6.93, 7.28, 7.099
7. D
8. 8, 637.4
9. 24/5 which is 4 ⅘; 4 pieces
10. $379.90
11. B and C
12. A: 84.38 B: 84.46 C: 84.40
13. 24 cubes
14. Tim has ⅓ more books than Raj. [4/3 = 1 ⅓ ]
15. 3.5
16. 4 ft.
17. 122 songs
18. 46 $units^3$

## Level 2

1. 5 units
2. B, C, and E
3. 21
4. Product 24; Term c and 2x; coefficient 2
5. 21 students
6. 174 girls
7. 120 tiles
8. A, B, and D
9.

| | 24 + 54 | 8 + 12 | 8 + 32 | 24 + 9 |
|---|---|---|---|---|
| 3(8 + 3) | | | | X |
| 6(4 + 9) | X | | | |
| 4(2 + 8) | | | X | |

10. 88 $in^2$
11. C
12. $1
13. 17.36
14. $4.00
15. A, C, and D

## Level 3

1. C
2. Any length between 3 ½ and 4 ½
3. ⅓ cup water
4. D
5. Red
6. Square
7. A
8. 480 $cm^3$
9. ¼ of a gallon
10. 120 deliveries
11. Without membership $442; With membership $420 ***
12. $90
13. B and F
14. Ben is using the expression before the discount is removed, Allison is combining terms to show how much you pay after the discount is applied. They are both correct.
15. 29/63
16. Two integers who have a sum of 1 (ex. -3 and 4; 2, -1)
17. Any fraction less that ¼ (ex ⅕)
18. a > b any a value must be larger than b value (ex. a= -1 and b= -4)

## Level 4

1. B
2. B and C
3. $2.94\ x\ 10^6$
4.

| | No solution | One solution | Many solutions |
|---|---|---|---|
| -6x + 2 = -6x + 2 | | | X |
| -8x + 7 = 8x + 7 | | x | |
| -9x + 8 = -8x + 9 | X | | |

5. $\sqrt{61}$ or 7.81
6. A and E
7. D
8. (0, -2)
9. g≥5
10. 50.24 $in^3$
11. C
12. 53.73 miles/second
13. 21 cards
14. B
15. -5x + 153
16. Any combination where a + b = -1 [ex. 4 + (-5)]
17. 2,000
18. Any equation where the slope (*m*) is > ½ and < 2
    (ex. y= ¾ x +2)

# Practice Test 2

## Level 1

1. 157, 000, 000 women
2. 4 ⅜ miles
3. $24 units^3$
4. a. 1/12 b. ⅚
5. 2 ⅛
6. 9.27 and 8.91
7. B
8. 6,257.8
9. 4 4/7; 4 pieces
10. $89.70
11. B and C
12. a. 82.25 b. 86.18 c. 82.84
13. 36 cubes
14. Jacob has ¼ more than Sal.
15. 3.6
16. 6 ft.
17. 109 songs
18. 56 $units^3$

## Level 2

1. 5 units
2. C and E
3. 17
4. Product 38; terms 4b and h; coefficient 4
5. 59 students
6. 167 girls
7. 114 tiles
8. A, B, and E
9.

| | 24 + 54 | 8 + 12 | 8 + 32 | 24 + 9 |
|---|---|---|---|---|
| 3(8 + 3) | | | | x |
| 6(4 + 9) | x | | | |
| 4(2 + 8) | | | x | |

10. 68 $in^2$
11. B
12. $1.50 per bottle
13. 17.36
14. $5.00
15. A, C, and D
16. 7.8
17. 2 ⅓ lbs.

# Level 3

1. D
2. Any number between 4 ½ and 8 ½
3. ½ cup of milk
4. A
5. Blue
6. Triangle
7. C
8. $224m^3$
9. ⅘ of a gallon
10. 45 deliveries
11. Without membership $243.75; with membership $183***
12. $57.50
13. B, C, and F
14. Brianna is using the expression before the discount is removed, Alexia is combining terms to show how much you pay after the discount is applied. They are both correct.
15. 1/24
16. Two integers who have a sum of 6 [ex. -2, 8]
17. Any fraction larger than ¼ but less that ¾ (ex. ½, ⅓ )
18. a < b (ex. a= -2 and b= 3)

## Level 4

1. C
2. A, C, and E
3. $1.63 \times 10^8$
4.

| | No solution | One solution | Many solutions |
|---|---|---|---|
| 6x + 11 = -6x + 11 | | x | |
| -8x + 7 = -8x + 7 | | | x |
| -3x + 4 = -4x + 3 | x | | |

5. $\sqrt{20}$ or 4.47
6. B and D
7. C
8. ( -1, 3)
9. $d \leq 7$
10. 17.96 $in^3$
11. A
12. 4 miles per second
13. 52 cards
14. D
15. 3x + 103
16. And combination where a + b = -1 (ex. a= -3 and b=2)
17. 3,000
18. An equation where the slope (*m*) is > ½ and < 2
    (ex. y= x -3)

Made in the USA
Monee, IL
05 November 2026

67387014R00149